PRAISE FOR *CONTACTEES*:

"Nick Redfern, the Brit with a knack for ferreting out all the dope on outrageous subjects, presents a revealing look at alien contact."

—Jim Marrs, author of *Alien Agenda*

"Nick Redfern, author of a wide variety of UFO books from the serious to the popular, has taken the bull by the horns and produced the, so far, definitive book on the subject."

—Andy Roberts, author of *Albion Dreaming*

"The disturbing fact, according to Redfern, is that a common thread tends to lead to the same, surprising conclusion in the bulk of cases. We are not alone, he posits, but the 'others' are not necessarily from Venus or the Pleiades. Where they do originate, and what it is that they want, I will leave to Redfern's eloquent explanation."

—Linda Godfrey, author of *Hunting the American Werewolf* and *The Beast of Bray Road*

"**Contactees** outlines a peculiar subculture that remains present in the fringes of Americana, and though well known to some, it has received its best treatment thanks to the efforts of Redfern."

—Micah Hanks, author of *Magic, Mysticism and the Molecule*

"A book that I couldn't set aside. Nick Redfern provides relevant information and pertinent minutiae about those now considered to be part of the UFO fringe."

—Adrian Wells, *The UFO Iconoclasts*

“Redfern gets down into the trenches to examine the band of eerily human-looking blond-haired visitors who early on expressed their concern about our warlike ways.”

—Timothy Green Beckley, *Bizarre Bazaar*

PRAISE FOR *MEMOIRS OF A MONSTER HUNTER*:

“**Memoirs of a Monster Hunter** is a wild and woolly five-year odyssey into the unknown, courtesy of one of the premier investigative researchers and authors in the field. Redfern’s adventures and often hilarious antics will leave you breathless.”

—Marie D. Jones, author of *2013* and *The Déjà vu Enigma*

“This is one of the best books I’ve read in years. Redfern sweeps you away on his personal adventure. Around the world, from romance, to ghastly beasts, to the cosmos, Redfern candidly shares the wonders of his young life.”

—Joshua P. Warren, author of *Pet Ghosts* and *How to Hunt Ghosts*

THE NASA CONSPIRACIES

The Truth Behind the **Moon Landings**, **Censored Photos**, and the **Face on Mars**

Nick Redfern

New Page Books
A Division of The Career Press, Inc.
Pompton Plains, N.J.

THE NASA CONSPIRACIES

EDITED AND TYPESET BY KARA KUMPEL

Cover design by Lucia Rossman/Digi Dog Design

Images on pages 9, 22, 30, 48, 53, 101, 104, 143, and 177 courtesy of NASA.

Image on page 44 courtesy of the White House.

Images on pages 171 and 185 courtesy of the author.

Printed in the U.S.A.

To order this title, please call toll-free 1-800-CAREER-1 (NJ and Canada: 201-848-0310) to order using VISA or MasterCard, or for further information on books from Career Press.

The Career Press, Inc.
220 West Parkway, Unit 12
Pompton Plains, NJ 07444
www.careerpress.com
www.newpagebooks.com

Library of Congress Cataloging-in-Publication Data

Redfern, Nicholas, 1964-

The NASA conspiracies : the truth behind the moon landings, censored photos, and the face on Mars / by Nick Redfern.

p. cm.

Includes bibliographical references and index.

ISBN 978-1-60163-149-7 ISBN 978-1-60163-679-9 (ebook) 1. Unidentified flying objects. 2. United States. National Aeronautics and Space Administration. 3. Conspiracy—United States. I. Title.

TL789.R367 2011

001.942--dc22

2010034227

Acknowledgments

I would like to take this opportunity to offer my very sincere thanks to the following: my literary agent, Lisa Hagan; Matthew Williams; Warwick Associates; and everyone at New Page Books, but especially Michael Pye, Laurie Kelly-Pye, Adam Schwartz, Kirsten Dalley, Gina Hoogerhyde, Kara Kumpel, and Diana Ghazzawi.

Contents

Introduction

On October 4, 1957, the entire Western world was well and truly shocked to its collective core when the former Soviet Union blasted into orbit its now legendary *Sputnik 1* satellite. And although the life of *Sputnik 1* was destined to be a manifestly short one—while decaying from its orbit only four days into 1958, it quickly ignited and was burned to cinders in the Earth's upper atmosphere—the propaganda value of its launch alone, at the height of the tension-filled Cold War, was near incalculable, and practically impossible to successfully trump.

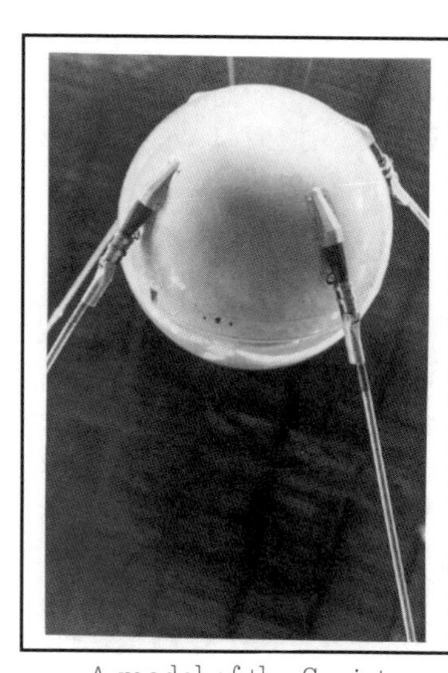
A model of the Soviet Sputnik 1 satellite.

The government of the United States of America, in particular, immediately felt both panicked and highly vulnerable due to the fact that the Russians had overwhelmingly beaten them in the first leg of the race to outer space. As a result of this setback for the United States, its government, its military, and the collective intelligence community of the CIA, NSA, and FBI, quickly recognized the dire need—scientifically, psychologically, and defensively—to catch up with what, at the time, was most certainly the biggest and largely unanticipated technological development within the Communist world.

Indeed, the U.S. Congress was so overwhelmingly worried, and utterly appalled by the Soviet's surprise leap into space, that it collectively demanded rapid, concerted, and unified moves on the part of the government as a whole to rectify the balance of power that had now been so drastically and quickly undermined. Under absolutely no circumstances at all, it was forcefully and logically argued, could the Soviet Union be allowed to gain a significant, or even a modest foothold in the previously uncharted domain of outer space; and particularly so if that very same domain was ever to become significantly militarized, as many figures within the scientific community, and within the Air Force and Army, suspected might one day very well occur—perhaps sooner than later.

President Dwight D. Eisenhower and his staff, astutely recognizing that the old, familiar world around them was changing rapidly, drastically, and in ways that had been largely unanticipated until now, embarked upon the first tentative and previously uncharted steps to try and rectify the situation, and set about balancing the precarious struggle for supremacy that still existed at the time between the powers of East and West. The dire need for a totally new body to deal with an equally new realm—namely, that of outer space—was clearly realized.

By the early months of 1958, the National Advisory Committee for Aeronautics (NACA) was well on its way to determining how, and under what particular circumstances an official organization of the U.S. government, unlike any that had ever previously existed, could take complete control, and carefully and capably manage the brave new world that outer space was offering humankind.

In April 1958, and as a direct result of NACA's growing vision, Eisenhower proudly stood before Congress and announced the ambitious establishment of what was to be originally known as the National Aeronautical and Space Agency. This was very good news, and precisely what the members of Congress dearly wished to hear. And, by the end of July 1958, the National Aeronautics and Space Act had been carefully formulated and approved at a presidential level. The new body, now to be known as the slightly reworded National Aeronautics and Space Administration—NASA, as it is known to one and all today—was duly born, and quickly initiated plans for the United States to play a decisive and leading role in outer space.

Since that now historic date, NASA has successfully placed countless satellites into Earth orbit; has blasted both men and women into space; has put a handful of brave astronauts onto the surface of our nearest neighbor, the Moon; has revamped and revolutionized off-planet travel with the Space Shuttle fleet; has sent unmanned probes to such planets as Saturn, Jupiter, Mars, and Venus; and has ensured that humankind is no longer tethered to planet Earth.

But that is not all: behind the scenes, there is a very different NASA; some might even say it's a darker and shadowy NASA. It is, as is about to become acutely apparent, a NASA that is seemingly populated to near bursting with stories of high-level cover-ups and secrets relative to:

- ✔ UFOs
- ✔ Flying saucers
- ✔ Alien life forms from faraway worlds
- ✔ Strange creatures
- ✔ Crashed extraterrestrial spacecraft
- ✔ Face-to-face encounters with the denizens of other worlds
- ✔ Dead aliens held in cryogenic storage
- ✔ Top Secret documents on lethal extraterrestrial viruses
- ✔ The notorious Face on Mars that many researchers of the puzzle believe was built millennia ago by a race of long-extinct Martians
- ✔ Classified and censored photographs of alien spaceships
- ✔ Shocking and sensational testimony from NASA's very own astronauts on their beliefs and personal sightings of a definitively unknown and alien nature

It is the strange, the fantastic, and the ominous world of the NASA conspiracies.

Chapter 1
Implications of the Alien Kind

Coincidentally or not, following the Soviet Union's successful launch of *Sputnik 1* in October 1957, there was a sudden and unsettling increase in the number of UFO sightings reported from within the confines of the United States of America. Whereas people of a skeptical nature might very well give much consideration to the possibility that many such reports were merely due to overexcitement, Cold War nerves, public hysteria and anxiety, and very understandable concern over the surprise Russian launch, other UFO-connected events could not be dismissed with such apparent ease and logic. A formerly secret FBI report of November 12, 1957, which has now been made available via the provisions of the Freedom of Information Act, makes that fact abundantly clear:

> Within the past two weeks reports have increased tremendously and some of the more serious have been described as follows: An object had landed in Nebraska with six people aboard, the persons had talked to a Nebraska farmer and then sped off into space; a fiery object was seen flashing across the southern skies from Albany, Georgia, to Miami, Florida; a Coast Guard cutter had sighted a huge object flying over the Gulf of

> Mexico; and persons in the Southwestern states while driving their cars have allegedly seen UFOs that caused the engines in their automobiles to stop.[1]

The FBI's special agents continued to diligently collate the strange and unearthly facts pertaining to what seemed to many of its personnel to be a near-cosmic invasion, and studiously briefed FBI Director J. Edgar Hoover on the nature of the expanding situation, as well as on the then-current response of the U.S. military to the vexing problem presented by the growing UFO presence:

> The Air Force is following these sightings closely and all reports are submitted to the Air Technical and Intelligence Center, at Wright-Patterson Air Force Base in Ohio where they are evaluated and analyzed. In the event any of the future reports appear to be authentic, the Air Force will immediately notify the Bureau, keeping in mind our particular interest in matters concerning espionage and sabotage.[2]

It is thought-provoking indeed to note that this undeniably dramatic upswing in UFO sightings, and even close encounters with alien entities, occurred in the immediate aftermath of the former Soviet Union's launch of its *Sputnik 1* satellite. Is it possible, perhaps, that the strange denizens of another world, or worlds, were secretly keeping a close watch on humankind's first attempts to break free of its terrestrial moorings? And, if so, was it those initial, hesitant steps outside of our own atmosphere that prompted such a flurry of concerned activity on the part of extraterrestrial visitors from far-away planets?

It may be important, and relevant, to note that by 1957, and throughout the course of little more than a decade, the human race had successfully developed atomic energy, had flattened two Japanese cities with atomic bombs, was working on advanced missile and rocket technology, and had finally left the moorings of the planet. In other words, it might very well be at this particular time in on our history, more than any other, when alien visitors from afar might begin to take a serious interest in us, and express deep concerns about us and our actions. Perhaps also, one might be inclined to speculate advanced alien civilizations of the type that were possibly secretly watching us in the late 1950s have undertaken such intense scrutiny and surveillance on countless occasions throughout the Universe. Particularly so when youthful, burgeoning civilizations take that giant and world-changing leap from being tied to their own planet, and when propeller-driven aircraft dropping bombs on the enemy are rapidly replaced by intercontinental missiles that have the ability to obliterate whole cities, countries, and cultures.

These controversial questions, issues, and speculation are made all the more provocative by the fact that only a couple of years later, at the dawning of the 1960s, when NASA's plans for outer space activity reached highly ambitious and groundbreaking levels, Donald N. Michael, who was then employed by the prestigious Brookings Institution, prepared a lengthy document on behalf of NASA's Committee on Long Range Studies, titled "Proposed Studies on the Implications of Peaceful Space Activities for Human Affairs, which was submitted to the House of Representatives in the 87th United States Congress on April 18, 1961. The creation, submission, and nature of the document in question proved to be a pivotal moment in the long and winding history of NASA and its relationship to all things of an unearthly and alien origin. The report was a truly significant one: in excess of 200 people, from all manner of disciplines that might have had a possible bearing upon the human race's future in the domain of outer space, were carefully consulted and interviewed at some depth.

In other words, this was a highly significant, unique project that, arguably, offered NASA's personnel a wealth of material in the form of expert advice, guidance, hypotheses, and recommendation on some of the major space-related issues of the day. The Brookings document makes for very notable reading for one specific reason: It compiles insights on the nature of extraterrestrial life, and the potentially dire implications for the entire human race if highly advanced alien cultures were one day discovered—or, on the other hand, if they were to discover us.

Even in its very earliest, formative years, NASA was deeply preoccupied with, and was focusing a great deal of its keen attention upon the theoretical notion of the human race making at least some form of contact with intelligent, extraterrestrial civilizations from far outside of our own solar system. The fact that the Brookings report, in no small part, dealt with the many and varied potentially thorny issues that might very well arise from close encounters of a definitively alien kind, must surely beg the significant question: Was NASA concerned, due to what had occurred directly after the Soviets' launch of *Sputnik 1* in 1957, that its own space missions might quickly provoke a wave of striking UFO encounters in and around the United States? Put another way, did that concern have any bearing, large or small, upon NASA's decision to commission the Brookings report in the first place? With those questions in mind, let us take a careful look at the relevant section of the now historic document itself: "Proposed Studies on the Implications of Peaceful Space Activities for Human Affairs."

It was one particular section of the report, titled "The Implications of a Discovery of Extraterrestrial Life," that generated so much commentary and interest at the time of its publication—even within the mainstream media that generally scoffed and sneered at any and all talk of friendly or hostile aliens visiting us from faraway worlds. But this was no amateurish, haphazard report prepared by wide-eyed science-fiction buffs or fanatical UFO devotees. No: The themes, the ideas, and even the warnings, relative to the many and varied potential outcomes that direct or even indirect interaction with aliens might provoke, were the results of some of the finest scientific minds within the United States of America applying their expertise to such matters.

Notably, and to the surprise of some within the mainstream American media, very few of those scientific minds consulted by Brookings were prepared to rule out the possibility that, one day, humankind might possibly find itself confronted by superior beings from a world perhaps very much like our own—or, conversely, one that was radically different from the planet we inhabit. Admittedly, the Brookings document clearly shows that there was major doubt and skepticism on the part of the scientific community of the day that literal face-to-face contact with E.T. would ever actually occur. Rather, it was near unanimously considered and concluded that radio would be the most likely medium by which we might one day finally obtain confirmation that the human race is not all alone in the Universe, after all.

Interestingly, it was also hypothesized within the pages of the Brookings document that perhaps ancient objects, devices, or structures left behind on the surface of the moon (or even on the surface of some of the nearby planets in our solar system) millennia ago by nonhuman intelligences might one day provide NASA with clues, and maybe even hard evidence, suggesting strongly that life out there had, at some point in our long and turbulent history, been far closer to home than we might previously have considered possible.

TOP SECRET

If the existence of alien life-forms was one day fully confirmed beyond any shadow of doubt, and then the determination was made that such a finding should be revealed to the general public and the media en masse, what would be the possible outcome? What would be the implications of such a revelation? Would worldwide chaos and fear quickly reign supreme? Might global social order completely and irreversibly teeter and wobble before finally collapsing and spectacularly imploding? Would there be amazement and bewilderment about our cosmic brothers and sisters and their intent, benign or otherwise, toward us? Could we find ourselves becoming overly reliant upon the presumed technological advances and scientific marvels that a race of beings centuries ahead of us might have to offer to the lowly human race? In other words, on this latter point, might our culture actually find itself swallowed whole by that of the near-omnipotent alien intelligences in our midst, to the point at which our present-day civilization and our familiar way of life one day becomes nothing more than the stuff of distorted memory, folklore, myth, and near-forgotten legend? These questions were vitally important and relevant ones to Brookings and to NASA.

Whatever the possible outcome, the Brookings Institution was confident that the answers to these and many more questions would be molded to a significant degree by the social, sociological, cultural, and religious beliefs of the general public all across the world, as well as by the similar beliefs, acceptances, and ideologies espoused by our elected leaders and religious authorities.

The plus side of all this speculation was the welcome scenario of the people of Earth finally uniting under one banner when faced with outright alien contact. In other words, following a revelation that aliens are among us and are here to stay, there might very well be a planet-wide push for us to see one and all as human beings, as citizens of a unified and a peaceful Earth, rather than—as we are now, I would strongly argue—a motley band of nations seemingly forever focusing upon conflict, one-upmanship, and national rivalry.

But, for all of that grand hypothesizing about what might happen should an announcement of alien contact be unleashed upon the world at large one day, some consideration was given within the pages of the Brookings document to the controversial scenario not of *when* the general public should be told that contact with alien intelligences had been successfully confirmed, but *if* that same general public should be told—ever. Of course, today, the inflammatory theory that elements of the U.S. government, the military, the intelligence community, and even NASA have indeed chosen to keep the public in the dark about their knowledge of alien visitations and UFOs is absolutely widespread. In view of this, perhaps we might argue convincingly that the Brookings report was not based solely upon mere speculation.

Unsurprisingly, the conclusions of the report made very big waves indeed within the American media of the day. As a perfect example of this, on December 15, 1960, none other than the *New York Times* devoted significant page space to the now historic and controversial report. The newspaper highlighted in part that NASA had been warned to prepare and ready itself for the discovery of advanced life-forms in outer space. The *Times* also carefully noted one of the most important and critical of all the issues raised: namely, the fact that the Brookings Institution report revealed that we, as a civilization, might suffer both adversely and significantly if confronted by a race of beings possessed of vastly superior intellect and highly advanced technologies.

Also very keen to comment upon the report from Brookings was one of the earliest, and certainly one of the most influential and respected, civilian UFO research groups within the United States: the National Investigations Committee on Aerial Phenomena (NICAP), which had been established in 1956 by a physicist and scientific visionary named Thomas Townsend Brown. In the December 1960/January 1961 issue of its in-house journal, *The UFO Investigator*, under the justifiable heading of "Space-Life Report Could Be Shock," NICAP echoed the words of the *New York Times* that the discovery of intelligent space beings in our midst could potentially have a severe and not necessarily positive effect on the entire global public mindset:

> The NASA warning of a possible shock to the public, from the revelation of more advanced civilizations, supports NICAP's previous arguments against AF [Air Force] secrecy about UFOs. All available information about UFOs should be given to the public now, so that we will be prepared for any eventuality.

It may very well be the case that NASA, too, had a deep desire, and also a pressing need, to be prepared for any and all eventualities—good, dire, foreseen, or otherwise. The Brookings report was not itself a classified, top-secret document; however, its stern warnings and observations to NASA about (a) the possible political and/or social repercussions that, theoretically, could be borne out of an announcement that aliens exist; (b) the potential disintegration of society that might follow in the wake of just such a life-changing revelation; and (c) the controversial issue of whether or not the news of a discovery of alien life should be withheld from the public, may very well have prompted the most senior and elite sources within NASA to secretly formulate plans to firmly bury, just about as far away as was humanly possible from inquisitive eyes and minds, any and all evidence of both extraterrestrial life-forms and UFOs.

Certainly, it is an absolute, undeniable fact that as the 1960s progressed, so did the many and varied claims and allegations linking NASA to high-level, UFO-dominated conspiracies and cover ups. That the space agency was steadfastly determined to demystify any and all assertions that it was deliberately hiding significant amounts of classified UFO data from the public, or, similarly, that it was sitting on sensational facts pertaining to the discovery of advanced, alien life-forms, might very well be perceived as hard evidence that NASA had chosen to take very careful heed of the Brookings report and its potentially world-changing opinions and paradigm-collapsing scenarios.

And, just before moving on to pastures new, it is worth noting the words of the late researcher Mac Tonnies, who made a valuable contribution to the debate on the Brookings document and its contents:

If our own history is any example, technologically robust civilizations inevitably subsume less sophisticated cultures, not merely by violently dismantling them, but by introducing a virulent strain of apathy. The infamous Brookings report to NASA, recommending that the discovery of extraterrestrial artifacts be covered up for fear of paralyzing research and development enterprises, stands as perhaps the most explicit elucidation of this idea. We appear to be interacting with an exceptionally patient intelligence which, despite its advantages over terrestrial science, seems limited by a steadfast refusal to make itself widely known. Whether this indicates a guiding morality or pragmatic necessity remains to be seen. Contrary to mainstream expectations, our visitors have opted for a much more gradual form of contact, evidenced both by the often theatrical nature of the apparent vehicles in our skies and by the behavior of the presumed occupants.[3]

Tonnies was not done:

I propose that this intelligence has played a significant role in occasionally hastening our species' development as well as keeping us in a periodic standby state, rendering us less likely to destroy ourselves. In a way, the human legacy has been scripted to conform to an alien template about which we know little or nothing. But the available historical, mythological, and experiential evidence tends to support a largely benevolent raison d'etre. Perhaps we're being groomed in preparation for our own singularity, after which the "others" could have no choice but to deal with us as equals.[4]

Chapter 2
From Mercury to Gemini

It cannot be denied that the overwhelming majority of NASA astronauts who have ventured into space have not spoken positively on, or imparted extraordinary data about UFOs, flying saucers, the theories and allegations that extraterrestrials may have clandestinely visited the Earth in our recent past, or that NASA possesses top-secret knowledge of such astounding matters. However, several most assuredly have made positive comments about UFOs and aliens, and, given the expert and legendary status of those making the claims, their comments make for highly significant and awe-inspiring reading.

We start with an absolute NASA hero, Gordon Cooper, who was one of the so-called Mercury Seven astronauts. In NASA's own words on Mercury:

> Implementation was initiated to establish a national manned space flight project, later named Project Mercury, on October 7, 1958. The life of Project Mercury was about 4 2/3 years, from the time of its official go-ahead to the completion of the 34-hour orbital mission of Astronaut [Gordon] Cooper.[1]

NASA's Mercury astronaut Gordon Cooper was a firm believer in UFOs.

Gordon Cooper claimed profound UFO encounters of his own, expressed a solid acceptance that aliens from outside of our solar system were among us, and even delivered noteworthy statements on the UFO controversy to prime and influential movers within the prestigious United Nations. Cooper first became immersed in the mysterious world of the unidentified flying object in 1951, when he had a series of intriguing sightings of flying saucer–style craft that spanned a period of several days, and while he was serving with the U.S. Air Force in what was then West Germany. It happened, said Cooper, as he and several other pilots were flying F-86 jets on a patrol over West Germany, and their attention was quickly drawn to what seemed to be a huge, high-flying convoy of circular-shaped aircraft. The total absence of wings, tails, propellers, or even jet engines, as well as their fantastic speeds and incredible maneuverability, allowed Cooper and his comrades to come to one conclusion, and one conclusion only: The unknown craft that had deigned them worthy of a personal viewing were nothing less than honest-to-goodness flying saucers from some other planet.

Astonishingly, for at least the next 72 hours, hundreds more such craft peppered the skies over West Germany, and, in the process, summarily outperformed the finest pilots and aircraft of the U.S. Air Force. Cooper quickly became convinced that the flying saucers were not the handiwork of the Russians, of the British, or even of his own government and military. Their point of origin had to be much further away—outer space, he could only conclude.

Noting astutely that the presence of alien spacecraft in our direct midst might very well have had some bearing on national security and defense-related issues, Cooper and his fellow pilots concluded that the wisest course of action was to quickly inform their immediate superiors of the incredible nature of their close encounters, of the many craft that they had all independently seen, and of the astonishing technology that those presumably piloting the UFOs clearly possessed, and were not at all afraid of demonstrating in the skies of Europe.

Those superiors were far from impressed by the accounts of Cooper and his friends, however, and outrageously waved each and every report aside, summarily and arrogantly dismissing them all as nothing more than examples of mistaken identity. What all of the trained, expert pilots had really seen, Cooper was boldly assured by senior Air Force personnel, were nothing more than high-flying seed-pods. Yes: seed-pods. I wonder, if highly skilled, long-term, expert fliers of the U.S. Air Force were wholly unable to differentiate between something as innocuous and down-to-earth as a seed-pod and something as fantastic, as futuristic, and as undoubtedly otherworldly as a flying saucer, then surely those same air crews should have been immediately grounded and subjected to a dizzying battery of psychological, physiological, and mental tests and evaluations?

They were not, however. Rather, the air crews were simply ordered to completely forget about what they had seen—or, rather, what they thought they had seen—and to continue with their regular patrols in the skies of West Germany, and to just concentrate on keeping a careful look out for far more down-to-earth things of a pressing nature, such as the military activities of the Russians. Someone within the senior ranks of the Air Force, it seems particularly safe to conclude, did not wish to draw any attention at all to an unearthly phenomenon that, as much as they preferred not to deal with it, seemingly had no intention of going away any time in the near future.

It was in 1957, however, that matters really heated up big time on the UFO front for the man who was destined, only a few years later, to become a leading player in NASA's Mercury space program. At the time in question, Cooper was just out of his 20s, had risen to the rank of captain, and had received a new assignment: to the Fighter-Section of the Experimental Flight Test Engineering Division at California's Edwards Air Force Base.

On one morning in the first week of May 1957, two colleagues and friends of Cooper—James Bittick and Jack Gettys—were situated out on a dry lake bed at Edwards AFB. The purpose of their presence at the lakebed was to take high-quality photographs of military aircraft as they came in to land at the base. In other words, it was just another surprise-free, routine day in the life of these two Air Force operatives. Or, at least, that's the way the day began. It did not stay surprise-free for very long, however.

In the course of the morning's events, Bittick and Gettys were both amazed and shocked to be confronted by nothing less than a *Day the Earth Stood Still*–style flying saucer that zoomed into view, hovered in the morning air for a while in a fashion very much like that of a helicopter, then carefully and briefly landed on the hot desert floor, only approximately 150 feet from where the shell-shocked pair could only stare in complete awe, before returning to the skies and vanishing in complete and eerie silence. Fortunately, however, as Bittick and Gettys were skilled photographers and cameramen, the pair had the presence of mind to secure both pictures and movie footage of the incredible scene that was fast unfolding before their disbelieving eyes. With such potentially priceless footage and imagery in hand, Cooper, after having been briefed on what had occurred, quickly placed a telephone call to the Pentagon to report on the startling nature of the morning's activities. Unsurprisingly, he was put through to a general, who made it very clear in stern tones that all of the relevant material evidence should be sent to the Pentagon without any form of delay whatsoever.

Cooper, astutely recognizing that questioning the orders of a full general was hardly the wisest move he could make, followed the orders down to the very last detail. There was, however, one matter of significance that Cooper noted:

> ...since nothing was said about not looking at the negatives before sending them east, that's what I did when they came back from the lab. I was amazed at what I saw. The quality was excellent, everything in focus, as one would expect from trained photographers. The object, shown close up, was a classic saucer, shiny silver and smooth—just as the cameraman had reported.[2]

Not everyone was in agreement with Cooper, Bittick, and Gettys on what the imagery showed, however. Major Robert F. Spence, of the Edwards Air Force Base Office of Information Services, stated of the controversial affair, when the salient details became known outside of official circles, that what Bittick and Gettys had actually seen was nothing stranger than a mere balloon that had been launched earlier that very same morning by the staff of a nearby "weather unit." And to reinforce the point that the U.S. Air Force was simply not prepared to hear any talk of UFOs and aliens under any circumstances at all, Spence carefully and forcefully added as a footnote:

> It is the opinion of the Air Force that any attempt to attribute anything unusual or mysterious to the incident is unwarranted and not supported by the facts.[3]

As for the seemingly invaluable film footage that Cooper said was taken on the fateful morning, and that reportedly showed what his two colleagues were certain was a genuine flying saucer–type craft, more than half a century later it has never publicly surfaced. And, essentially, that is where matters continue to rest to this very day. UFO vs. balloon: The jury is still controversially out on that particular issue. Cooper's involvement in things both flying and saucer-shaped was very far from being over, however. In fact, it might be correct to say that it was just getting started.

CLASSIFIED

Sir Eric Matthew Gairy held the position of Premier of the island of Grenada from 1967 to 1979. In 1977, two years before he finally left office, Gairy began to enthusiastically lobby the United Nations to create an agency, office, or department designed to "collate, coordinate and corroborate information" on UFOs and extraterrestrial life.[4]

In October 1977, Gairy said in a memorable statement to the United Nations on the specific subject of UFOs:

> I think it is accepted that these things do exist. I think we now want to know the nature, the origin and the intent of these saucers. Some people think they have come to do good. Some think they have come to dominate human beings.[5]

Significantly, Gairy also cited several cases in which a number of "aircraft have been put out of commission, but not destroyed, after attacking saucers." According to Gairy's personal and particular, comforting train of thought on this specific aspect of the UFO phenomenon:

> That confirms my thought on their positive intent: I believe they are coming here to help mankind because man is so self-destructive.[6]

As a direct result of Gairy's unique prompting to the United Nations, on November 9, 1978, NASA astronaut Gordon Cooper submitted memorable words of encouragement and support in the form of a letter to Ambassador Griffith, Mission of Grenada to the United Nations. Cooper was unequivocal and concise, as his words, recorded in official Department of State files, clearly demonstrated. He stressed that the establishment of any such organization along the lines suggested by Gairy would require it to have the ability, scope, and staffing to capably and scientifically study the UFO subject to a truly significant and previously unparalleled degree.

In specific terms of the aliens themselves, which Cooper solidly accepted were now visiting us, he told the United Nations that:

> We may first have to show them that we have learned to resolve our problems by peaceful means, rather than warfare, before we are accepted as fully qualified universal team members. This acceptance would have tremendous possibilities of advancing our world in all areas. Certainly then it would seem that the U.N. has a vested interest in handling this subject properly and expeditiously.[7]

Despite the very best efforts of Cooper, Sir Eric Gairy's ambitious *X-Files*–style ideas never really amounted to anything of real, meaningful significance. Cooper was far from being done yet, however, when it came to trying to alert the human race to the unearthly UFO presence that he believed was directly among us, and had been for decades, and perhaps even longer. As evidence of this, in 1980, Cooper made a number of notable, public statements on the UFO controversy. When questioned in that year by *Omni* magazine on the subject of his 1951 encounter, Cooper outright confirmed and reinforced its reality, adding in truly significant tones, given his standing and experience with the U.S. Air Force, and later with NASA, that:

> From my association with aircraft and spacecraft, I think I have a pretty good idea of what everyone on this planet has and their performance capabilities, and I'm sure some of the UFOs at least are not from anywhere on Earth.[8]

Then, in 1985, when UFOs were once again briefly on the agenda of the United Nations, Cooper offered the following significant words that were recorded for posterity by the Department of State:

> I believe that these extraterrestrial vehicles and their crews are visiting this planet from other planets, which are a little more technically advanced than we are on Earth.[9]

Once again, however, despite the important and almost unparalleled fact that none other than a greatly respected, retired NASA astronaut and American hero was personally endorsing and highlighting the theory that some UFOs may possibly represent alien spacecraft, the United Nations failed to act to any meaningful or worthwhile degree. To some players within the public UFO research community at the time, this admittedly disappointing outcome was perceived as prime evidence that the United Nations was dissuaded from pressing ahead with its own UFO study program, by even more powerful and shadowy forces buried deep within the realm of international officialdom. For the open-minded skeptics and the outright debunkers, however, the United Nation's decision not to establish an official body to investigate UFOs, aliens, and flying saucers was perceived as hard evidence that a conclusion had been reached that there was nothing worth investigating in the first place.

Cooper, meanwhile, continued to vigorously stand by his claims of UFO encounters in the skies of West Germany in 1951. And, whenever the opportunity arose, he never failed to champion the Edwards Air Force Base missing-UFO-film affair of May 1957. Cooper died at the age of 77 in October 2004, a firm believer to the very end that UFOs exist, and that they're not from any neighborhood nearby.

CONFIDENTIAL

Gordon Cooper wasn't the only one of NASA's original seven Mercury astronauts to have reported a personal UFO encounter: Donald Kent "Deke" Slayton did as well. While flying in 1951, Slayton said, he had his first (and, as history would demonstrate, his only) close encounter with what some have said was an intelligently controlled vehicle that originated within the depths of another galaxy.

Slayton explained the fascinating facts:

> ...I had just come out of a spin at around 10,000 feet over the Mississippi River...I was heading back to Holman Field when all of a sudden I saw this white object about my altitude, at one o'clock.... My first thought was that it looked like a kite. But logic said nobody's flying a kite at this altitude. So I started kind of watching it to see what it was.... The closer I got, the more it looked like a weather balloon, and I'm thinking, that's what it's gotta be. Then I flew past it a little high, about a thousand feet off. It still looked like a 3-foot-diameter weather balloon to me.[10]

Due to the fact that, initially at least, he was pretty much convinced he had seen nothing stranger than a small weather balloon, Slayton chose to remain silent on the incident. That is, until only a few days later when, after casually telling his boss about the event in question, Slayton was told quickly and decisively: "Get your ass over to Intelligence in the morning and give them a briefing." Having duly done so, Slayton was quietly advised by his Air Force interviewers that on the same day as his own sighting, a local company was flying high-altitude research balloons in the same area. While doing so, the balloon team had also seen a mysterious flying object that hovered in the skies before taking off "like hell."[11]

Slayton, always the balanced and careful thinker, preferred not to speculate on the affair, and could only conclude that:

> My position is, I don't know what it was: it was unidentified.... It's still an open question to me.[12]

It is worth noting, contrary to what many might initially assume, that UFOs are certainly not always reported as being gigantic in size. Indeed, many qualified observers have reported seeing very small vehicles of unknown origins in our skies—not at all unlike the one encountered by NASA astronaut Slayton.

For example, FBI-originated UFO files of 1952 refer to the sighting of a UFO by a lieutenant commander with the U.S. Navy, who had seen an approximately 4-foot-diameter UFO hovering in the sky in March 1952, near his Chicago, Illinois, home. Similarly, additional FBI files of the same month refer to yet another sighting of a small UFO in the vicinity of Chicago. In this particular case, the FBI noted that the witness "described the disc as approximately 6 feet in diameter, circular, white in color with a bluish tinge. The disc, he said, appeared to have been constructed out of a metal similar to aluminum."[13]

Although I am personally convinced that some UFOs are indeed vehicles piloted and controlled by beings from other worlds and realms of existence, perhaps these particular types of reports—of very small UFOs, such as that seen by Mercury astronaut "Deke" Slayton in 1951—are an indication that the intelligences behind the UFO phenomenon are not beyond utilizing small, unmanned, remotely piloted vehicles, perhaps undertaking reconnaissance missions deemed far too risky for a fully crewed, large vessel. Granted, this is nothing more than speculation on my part, but it is a point worth considering and keeping in mind, and which may go some way toward explaining the presence of such compact UFOs.

TOP SECRET

After Mercury, NASA's later manned space program, known as Gemini, was also destined to become forever linked with the ideas of extraterrestrials and flying saucers. This is what NASA has to say about Gemini:

> The National Aeronautics and Space Administration announced December 7, 1961, a plan to extend the existing manned space flight program by development of a two-man spacecraft. The program was officially designated Gemini on January 3, 1962.[14]

NASA's Gemini program has long been linked with the UFO mystery.

The Gemini UFO connection came from a strange story that surfaced in the pages of a report prepared by what became known as the Condon Committee—an informal title for the University of Colorado's UFO project that ran from 1966 to 1968, under the direction and control of physicist Edward U. Condon. As prime evidence of the link between NASA's Gemini project and the mystery of unidentified flying objects, take careful note of the following section of the official documentation generated during the course of the Condon Committee's work, which focused upon three NASA/UFO events that the committee came to believe had not been resolved to a wholly satisfactory degree. Those particular events were recorded by the committee in the following fashion:

> Gemini 4, astronaut McDivitt, observation of a cylindrical object with a protuberance. Gemini 4, astronaut McDivitt, observation of a moving bright light at a higher level than the Gemini spacecraft. Gemini 7, astronaut Borman saw what he referred to as a "bogey" flying in formation with the spacecraft. Gemini 4, cylindrical object with protuberance. Astronaut McDivitt described seeing at 3:00 CST, on 4 June 1965, a cylindrical object that appeared to have arms sticking out, a description suggesting a spacecraft with an antenna. I had a conversation with astronaut McDivitt on 3 October 1967, about this sighting...McDivitt saw a cylindrical shaped object with an antenna like extension... it was not possible to estimate its distance but it did have angular extension; that is it did not appear as a "point."[15]

We may never really know if McDivitt's sighting was of a genuine UFO, or, as some later concluded, of the second stage booster of his own Titan II rocket. There is, however, a conspiratorial footnote to this potentially important affair. A heavily censored, formerly classified FBI document of September 2, 1965, gives every indication that a source within NASA had access to incredible UFO data that had a direct and distinct bearing upon the *Gemini 4* mission. Furthermore, recorded the FBI, this same individual was reportedly intent on covertly forwarding the data on to persons without any form of official clearance, or ties to NASA, whatsoever.

According to the presently available FBI memorandum, the source, whose name remains fully removed from the official record, had informed special agents at the FBI office at Pittsburgh, Pennsylvania, that two people—one of whom was described by the FBI as being a graduate student at Pittsburgh University—were "acquainted with a NASA employee [name censored by the FBI], and have stated that he furnishes them information by mail about unidentified flying objects (UFO) which he obtains from NASA files. The source believes that the information may be classified."[16]

The FBI documentation on this matter continued in remarkable tones:

> The source said, for example, that [censored] had seen a motion picture film showing a missile separation and an UFO appearing on the screen. Prior to the flight of Gemini 4 [censored] said to watch out for something interesting because the space ship had devices aboard to detect UFOs.[17]

Solid proof that the information divulged by the NASA employee was considered to be somewhat sensitive in both its nature and its content is clearly demonstrated by a further extract from the document that is still contained in the FBI's archives:

> [The NASA source] posts his letters in a mail box away from NASA and puts hairs in the glue of the envelope so that the addressee can determine if the envelope was opened. This source stated he had no reason to believe that the information was going to any foreign power.[18]

Although admittedly brief in nature, and heavily redacted by the FBI prior to its declassification, the content of this particularly eye-opening document raises a number of important issues, such as:

- Who, precisely, was the NASA whistleblower?
- From where, exactly, was he securing his seemingly classified UFO information?

- What of the allegations of the NASA "Deep Throat" to the effect that *Gemini 4* was equipped with certain, unspecified devices to monitor for UFO activity while in space?

Whatever the truth of the *Gemini 4* UFO story, these curious revelations that originated with the FBI can only leave us with a strong, nagging suspicion that we may not have heard the last of UFOs and NASA's Gemini program—not by a long shot.

Chapter 3
Crash at Kecksburg

Late in the afternoon of December 9, 1965, after having first been seen high in the sky as a glowing ball of fire that crossed a number of U.S. states and even parts of Canada, an object of distinctly unknown origins—perhaps a literal alien spaceship from some distant world—slammed into the ground in deep, shadowy woods near the small Pennsylvania town of Kecksburg, which is situated approximately 30 miles southeast of the city of Pittsburgh.

In the immediate aftermath of the controversial event, dark tales of military personnel descending upon the scene like veritable flies, of witnesses being silenced, of a UFO, of alien bodies having been found in the woods, and of conspiracies of a near cosmic kind abounded—and, to this date, continue to do so. Fortunately, none of the reported intimidation of witnesses has prevented the vast majority of them from telling their startling stories of that long gone day. And NASA, as some researchers of the affair have concluded, continues to stonewall at every opportunity on its reportedly secret knowledge of what did or did not happen on that memorable day in those darkened, Pennsylvania woods.

Stan Gordon, a dedicated and intrepid researcher and writer on a wide range of anomalies, has done more than just about anyone else to try and unravel the many and varied complexities of the Kecksburg affair. For example, during the course of his research, he uncovered the account of a Bill Bulebush, who was tuning his CB radio in his car that 1965 afternoon when he looked up to see the fiery object moving from Norvelt toward the mountain near Laurelville. "Bulebush said the object appeared to hesitate over the Laurelville area," noted Gordon, adding that the craft "then made a turn and then began to travel northeast toward Kecksburg, where he saw it descending.... It appears that the object may have been slowing down before its descent into the woods near Kecksburg, only miles away."[1] Clearly, this is important testimony: If the unknown object that came down at Kecksburg possessed the ability to actually slow down its movements in midair, then it can hardly have been a chunk of space-rock. Rather, the words of Bill Bulebush strongly suggest the object was under some form of intelligent control.

Not surprisingly, when one takes into consideration the fact that this was a relatively small and very close-knit community, the Kecksburg affair was immediately thereafter major news at a local level. As prime evidence of this, at 9 p.m. on the same night that all hell was breaking loose around Kecksburg, the local radio station, KDKA, broadcast a statement to its listeners, many of whom were keenly and fully aware that something very weird was well and truly afoot in the immediate area.

The station excitedly reported that an unidentified flash of orange-colored light in the sky had prompted numerous people throughout Ohio, Indiana, Michigan, Pennsylvania, and the Ontario province of Canada to contact the military and emergency services to report their own sightings of the mysterious intruder. As a result, elements of the Air Force took careful initial steps to try and resolve the mystery, but, publicly at least, could find no evidence whatsoever that either an aircraft or a missile was the definitive culprit.

Meanwhile, added KDKA, the Federal Aviation Authority had offered the seemingly plausible suggestion that the unknown object could have been one of two things: a meteor burning up in the Earth's atmosphere, or the charred remains of a space vehicle—whether American or Russian—that was viewed while it was reentering the atmosphere. Perhaps not surprisingly, staff at local missile bases chose not to make immediate comments, as the radio station noted: "They say that they will issue a statement within a half hour. Many persons in the Greensburg area saw the phenomena. They are investigating. The Oakdale Missile Master was contacted. They said a release is coming up. Speculation also is that an Army missile went astray."[2]

Significantly, there was deep debate and speculation on the part of many of the townsfolk of Kecksburg that what had really come down near their little town was nothing less than a craft from E.T. itself. Certainly, on the day and night at issue, the whole town was a hotbed of chaotic activity and wild rumor. Local firefighters, media journalists, and the radio news director from WHJB Radio all quickly descended upon the scene. They described seeing a sizeable and significant military presence, either on the outskirts of or deep inside the dense woods. Something strange, surely, was afoot. And just maybe it was something alien too.

A significant body of individuals said that, while trying to determine what had happened in the woods, they were confronted by numerous uniformed personnel brandishing firearms, no less. Others, meanwhile, stated unequivocally that they saw an impressively sized metallic-looking device at the site that was subsequently removed by the military. Given the fact that such cloak-and-dagger activity was afoot, it is no surprise at all that only one day later, the *Tribune-Review* newspaper, which covered the area in question, gave the sensational story pride of place in its pages: "Unidentified Flying Object Falls near Kecksburg—Army Ropes off Area."

This was all in stark contrast to the far more down-to-earth assertions of the U.S. Air Force's UFO investigative operation, Project Blue Book. A spokesperson for Blue Book said that no U.S. space debris, at all, had been reported reentering the Earth's atmosphere on the day in question, and so could not possibly have accounted for the reported incident. But the sensational idea that a spacecraft from another world had crashed, or had landed, was dismissed as being totally absurd, and without any foundation in reality. In addition, stated the Blue Book representative, even though a failed space probe launched by the Soviet Union, classified as Cosmos 96, did reenter the Earth's atmosphere over Canada on that morning, its particular trajectory meant that it could not have played any sort of role—meaningful, tangential, or otherwise—in the strange events at Kecksburg.

The Blue Book official was also very keen to stress to the local media that the only official presence at the presumed site of the impact was a three-person team that had been dispatched by the Air Force from a radar installation located near the city of Pittsburgh. Their particular job, it was explained to anyone and everyone who might have been willing to listen, was to search for any and all evidence of what had possibly come down. Officially, however, nothing was ever found. And the only thing the Air Force could suggest, by way of some form of explanation, was that nothing stranger than an everyday meteorite was the real culprit behind the controversy.

Witness testimony, however, strongly suggested otherwise. Stan Gordon revealed that in 1990 he had been contacted by a man who confided in him extensive and extraordinary data that appeared to be directly relevant to the Kecksburg event. At the time in question, the man was attached to an Air Force security team stationed at Lockburne Air Force Base, near Columbus, Ohio—a team that, Gordon's informant told him, stood watchful and careful guard over the Kecksburg device in a secure aircraft hangar during the early hours of December 10, 1965, after it had been secretly transferred from its impact point in Kecksburg, Pennsylvania. Security at the base, Gordon was further informed, was even more stringent than when President John F. Kennedy had made a visit there some years earlier. So the man's revelations went, the object stayed at Lockburne for perhaps only a matter of hours at the most. Afterward, it was reportedly clandestinely transferred to Wright-Patterson Air Force Base near Dayton, Ohio, which just happened to be home to the Air Force's Foreign Technology Division (FTD).

CLASSIFIED

The FTD would have been the ideal location from which any vehicle or device of non-U.S. origins could be carefully examined in depth and amid great secrecy by some of the military's finest scientific minds. And it may not just have been a craft of unknown origin that was secretly taken to Wright-Patterson Air Force Base either: some have suggested that alien bodies—and, astonishingly, perhaps even *live* aliens—may very possibly have been found and retrieved at the site of the impact.

Possibly of deep relevance to this particularly controversial aspect of the story is the startling account of Don Sebastian, a resident of Johnstown at the time, who was in the Kecksburg area visiting friends when the incredible news about the Kecksburg crash broke over the airwaves of the area. Notably, an attempt by Sebastian and his friends to secure access to what was presumed to be the crash site resulted in them hearing a blood-curdling scream, quite unlike any other that they had ever heard in their lives before, which echoed around the shadowy woods. Sebastian didn't have to think twice about getting the hell out of there, and at high speed. Maybe he didn't even need to think once.

CONFIDENTIAL

Stan Gordon has heard similar stories of alien creatures having been reportedly found at the scene of the Kecksburg incident. One particular witness, who worked at Wright-Patterson Air Force Base at the time, and whom Gordon prefers to refer to only as Myron, saw "a body lying on a workbench in that same room...which he estimated to be approximately 4 to 5 feet tall and would weigh about 80 pounds," and that was described as being "lizard-like."[3]

TOP SECRET

The mystery of what did or did not happen at Kecksburg on that 1965 day continues to survive and thrive nearly half a century after it occurred. The event was accorded a huge amount of publicity, and even notoriety, one might be inclined to say, in 2003, when none other than the Sci-Fi Channel (now known as SyFy) decided to get involved in trying to finally unravel the complexities of the near 50-year-old mystery from beyond the cosmos. As a direct result of the Sci-Fi Channel's actions, it wouldn't be very long at all before NASA found itself up to its collective neck in the Kecksburg quagmire too.

One area quickly addressed by the Sci-Fi Channel when it got hot on the trail of the story was whether or not the Kecksburg affair could have been linked, in some not entirely clear fashion, with the reentry into the Earth's atmosphere of the Soviet Union's *Cosmos 96* satellite that did indeed occur on that same day. Certainly, the satellite was somewhat acorn-shaped in appearance, which, in the eyes and minds of some UFO researchers and a number of eyewitnesses at least, did offer a certain degree of credence and merit to that particular theory. There is, however, a very big problem with the notion that the U.S. military clandestinely retrieved the Russian vehicle from the woods outside of Kecksburg: *Cosmos 96* actually crashed on Canadian territory a number of hours before the Kecksburg events even kicked off. And to reinforce this point, in a 2003 interview, the Chief Scientist for Orbital Debris at the NASA Johnson Space Center, Nicholas L. Johnson, said forthrightly: "I can tell you categorically that there is no way that any debris from Cosmos 96 could have landed in Pennsylvania anywhere around 4:45 p.m. That's an absolute. Orbital mechanics is very strict."[4]

In addition to addressing the potential *Cosmos 96* connection, there was a determined push by the Sci-Fi Channel to encourage NASA to release any and all official documentation on the Kecksburg affair that might conceivably be contained within its impressive, bulging, decades-old archives. As a direct result of this push for information, in November 2003 more than three dozen pages of Kecksburg-based material were finally released into the public domain by NASA personnel. Unfortunately, and despite initial encouraging hopes to the contrary, the papers at issue failed to shed any form of real, meaningful light on what did or did not actually occur on the afternoon of December 9, 1965. A similar search undertaken by NASA personnel in the summer of 2006 turned up the same lack of information.

CLASSIFIED

It was, therefore, just about the right time to take things to a whole new and unique level. On March 27, 2007, United States Judge Emmet G. Sullivan signed off on a civil action against NASA that had been initiated by a sleuth-like Kecksburg researcher, namely the Director of Investigations of the Coalition for Freedom of Information, Leslie Kean. The purpose of this new approach was to finally try and secure public access to any and all files, records, documents, and memoranda held by NASA on the Kecksburg crash, with the hope of eventually revealing what it really was that came down in those Pennsylvania woods all those years ago—and what NASA really knew of the tumultuous affair. The ultimate battle between NASA and the public UFO research community, as it related to the crash at Kecksburg at least, was now decisively underway.

The civil action was filled with page after page of nearly mind-numbing legal speak, but did highlight a number of important points, including:

- Contradictory statements that NASA had made relative to its files (or, more correctly, its pronounced lack of files) on the Kecksburg affair.

- Kean's deep-seated concerns about the less-than-satisfactory ways in which NASA had conducted its searches for relevant documentation on the events of December 9, 1965.
- What appeared to many researchers to be highly suspicious actions on the part of NASA to try and thwart any and all attempts to successfully open up the Kecksburg can of worms.

The pro-UFO research community was still very far from being finished, however, and continued valiantly to do battle with NASA. In November 2009, Leslie Kean revealed the latest in this seemingly neverending saga, and focused her attention upon NASA's curiously absent documentation on Kecksburg and the issue of files that—deliberately or mistakenly—NASA had possibly destroyed. Kean herself admitted, when summing up the strange and unearthly situation still facing her: "Without additional, very extensive work, we'll never know the answers, and even with the work, we still might never know."[5]

CONFIDENTIAL

Despite the fact that, today, the Kecksburg affair still continues to languish in a twilight realm that some see as being dominated by deep conspiracy, a crashed UFO and its dead alien crew, and others view as the combined results of bureaucracy and a meteorite, Kean said, after the confrontation with NASA was finally at an end that she was still convinced something crashed at Kecksburg—and it was not the former Soviet Union's *Cosmos 96*, nor any other form of Russian device. Kean also considered it unlikely that the Kecksburg affair could be successfully explained as a secret American space capsule of some sort.

For some, Kecksburg was, and certainly still is, much ado about not much at all—at the very most, a bit of space debris entering the Earth's atmosphere. For others, such as Stan Gordon and Leslie Kean, however, Kecksburg is the one case that, perhaps more than any other, might very well prove to the world at large that NASA secretly has in its possession demonstrable evidence that extraterrestrial life has visited our world, has crashed here, and may even have died here too. Whatever the strange truth of the matter may one day prove to be, it seems highly unlikely that the controversy surrounding the case is destined to go away anytime in the foreseeable future.

The final word, for now, goes to a highly prestigious source, John Podesta, who was President Bill Clinton's former chief of staff, and a member of the 1997 Moynihan Commission on Protecting and Reducing Government Secrecy. He specifically said of the Keeksburg saga of December 1965, and of Leslie Kean's valiant attempts to secure the facts, whatever they might be and wherever they may lead, from NASA:

> It's time to find out what the truth really is that's out there. We ought to do it because it's right; we ought to do it because the American people quite frankly can handle the truth; and we ought to do it because it's the law.[6]

Chapter 4
Apollo: Flights of Fancy?

On September 12, 1962, the late President John F. Kennedy delivered what was destined to become a groundbreaking speech at Rice Stadium, at Rice University in Houston, Texas, that set the scene for NASA's historic *Apollo 11* landing on the surface of the moon in July of 1969—an event that whole swathes of the American population, today, solidly believe or strongly suspect was nothing more than an audacious hoax on the entire world, as will become clear very soon. But, before getting to the various claims of fakery, obfuscation, and stage-managed trickery, it is important to take keen note of certain integral aspects of President Kennedy's now legendary speech. Their historical relevance in the controversy surrounding the moon landings is paramount.

Kennedy told the entranced audience on that 1962 day:

> The exploration of space will go ahead, whether we join in it or not, and it is one of the great adventures of all time, and no nation which expects to be the leader of other nations can expect to stay behind in the race for space.... We set sail on this new sea because there is new knowledge to be gained, and new rights to be won, and they must be won and used for the progress of all people....

> We choose to go to the moon. We choose to go to the moon in this decade and do the other things, not because they are easy, but because they are hard, because that goal will serve to organize and measure the best of our energies and skills, because that challenge is one that we are willing to accept, one we are unwilling to postpone, and one which we intend to win....

President John F. Kennedy spoke at Rice Stadium, Houston, Texas, in 1962. The subject: NASA's plans to send men to the moon.

With those memorable and sterling words issued forth by the ultimately doomed Commander in Chief, a whole nation was galvanized, and NASA was duly overjoyed. But did NASA really win the race to the moon, or was the whole Apollo program nothing more than a gigantic ruse, one that still remains relatively intact to this very day?

TOP SECRET

It was on the afternoon of July 20, 1969, at 4.17 p.m. that history was made. On that day, and while then–U.S. President Richard M. Nixon anxiously viewed the now-legendary proceedings from within the heart of the White House's Oval Office, humankind took its very first, tentative steps onto the surface of our nearest cosmic neighbor: the moon. Launched into the depths of space only four days earlier from the Kennedy Space Center at Merritt Island, Florida, the *Apollo 11* crew of Neil Alden Armstrong, Edwin Eugene "Buzz" Aldrin, and Michael Collins collectively represented the pinnacle of a nearly decade-long program to achieve what many had previously perceived to be little more than outrageous science fiction and fantasy—and which many still assert to this day *is* nothing less than outrageous science-fiction and fantasy.

Fully 43 years before the Apollo 11 landing, Dr. Lee De Forest, a genius in the field of electronics, stated that it was sheer folly to even seriously contemplate the idea of sending astronauts to the moon. Likening such an idea to a Jules Verne fantasy novel, he was sure that such an event would "never occur."[1]

History has shown that, for most people at least, the good doctor was drastically wrong in his loud proclamation. Somewhat ironically, the late actor DeForest Kelley, who portrayed Dr. Leonard "Bones" McCoy on the 1960s television series *Star Trek*—which successfully brought the concept of outer space exploits and adventures of a very alien kind to the mainstream public at large—was himself named after Dr. Lee De Forest.

CLASSIFIED

The historic events of July 1969 were followed by five more manned missions to the moon, involving no less than a dozen astronauts successfully walking on its surface, which proceeded to captivate and enthrall the entire world for years. Beyond any shadow of doubt, the Apollo landings are, today, more than 40 years after they began in earnest, perceived as being defining moments in humankind's history, and, collectively, one of the finest and most lasting legacies of the administration of President John F. Kennedy.

But were the manned Apollo missions to the moon precisely what they seemed to represent, and what NASA loves to tell us they represent? Despite the very vocal, angry, and frustration-driven protestations of NASA and the collective Apollo astronauts who set foot on the moon, not everyone seems to think so. Certainly, there is a very deep suspicion, and even an overwhelming, cynical acceptance on the part of whole swathes of the American population that the Apollo moon landings were nothing less than outrageous trickery, the purpose of which was to instill in the minds of the Soviet hierarchy the firm belief that American technology was far in excess of anything with which the best minds of the Russian scientific community could ever hope to compete.

As the admittedly controversial theory goes, the Soviet Union had beaten the United States in the race to place a satellite in Earth orbit with *Sputnik 1* on October 4, 1957. Worse still, the Soviets were also the first people to place a man in orbit around the Earth: namely, Yuri Alekseyevich Gagarin, who achieved his legendary status on April 12,

1961. As a result of these two significant achievements by the Soviets, the conspiracy theorists say, there was simply no way whatsoever that the United States of America could afford to lose what was then perceived as being the most important race of all: to put a man, or men, on the surface of the moon. One way or another, NASA simply had to be the winner in this particular cosmic battle. Or, if NASA couldn't outright win, then it would have to do the near unthinkable: It would be forced to convince the world it really had won, via nothing less than total deception.

Are such controversial assertions simply the crazed ravings and rants of wild-eyed conspiracy theorists that see cover-ups and dark secrets behind every door? Or, astonishingly, could there really be a degree of truth to some of these jaw-dropping assertions and accusations?

CONFIDENTIAL

In 1969, the renowned American journalist and novelist Norman Mailer wrote, with respect to the first moon, that:

> The event was so removed, however, so unreal, that no objective correlative existed to prove it had not been an event staged in a television studio—the greatest con of the century—and indeed a good mind, product of the iniquities, treacheries, gold, passions, invention, deception, and rich worldly stink of the Renaissance could hardly deny that the event if bogus was as great a creation in mass hoodwinking, deception, and legerdemain as the true ascent was in discipline and technology. Indeed, conceive of the genius of such a conspiracy. It would take criminals and confidence men mightier, more trustworthy, and more resourceful than anything in this century or the ones before. Merely to conceive of such men was the surest way to know the event was not staged.[2]

Not everyone agrees, however. NASA authority James Oberg said both accurately and with much justification that: "There are entire subcultures within the U.S., and substantial cultures around the world, that strongly believe the [*Apollo 11*] landing was faked." Notably, Oberg also revealed that within the school system on the island of Cuba, the teaching of the moon landings being nothing more than audacious hoaxes was still very much an ongoing issue.[3]

Moreover, a Gallup poll of 1999 demonstrated that 6 percent of all Americans questioned had expressed at least some suspicions that the Apollo missions did not occur in precisely the fashion that NASA, and its Apollo astronauts, have consistently and loudly asserted. At first glance, at least, 6 percent might not sound like a lot at all. That is, until the realization hits home that this equates to approximately 18 million Americans. This, of course, is hardly an insubstantial number of people. In fact, it's precisely the opposite.

TOP SECRET

Under what particular circumstances have such controversial doubts crept into the collective mind of modern-day society? Why is there an acceptance that NASA has, for more than 40 years, engaged in a massive deception that makes the combined Watergate, Iran-Contra, and Weapons of Mass Destruction controversies pale in comparison? To answer those particularly thorny questions, we have to go back to 1974, when a truly controversial, albeit small and privately published and circulated book—*We Never Went to the Moon: America's Thirty Billion Dollar Swindle*, written by the late Bill Kaysing—began making waves within the media, among the public, and, of course, within much of NASA itself.

Kaysing strongly proclaimed that in the immediate years before the moon landings and directly afterward, the stark reality of the situation was that NASA simply did not possess the scientific capabilities to successfully send teams of astronauts to the moon, nevermind to successfully and safely return them to Earth afterward, and with the priceless film footage, photographs, and health of the astronauts wholly intact in the process.

Kaysing's primary arguments against the moon landings have since been championed and elaborated upon by numerous other conspiracy theorists since they first surfaced in 1974, but, for the most part they were focused on a couple of key questions:

- Why was there a total lack of visible stars in many of the pictures said to have been taken by the Apollo astronauts while they were on the moon?
- How was it possible that the American flag—famously planted by the astronauts on the surface of the moon on that life-changing day—seemed to almost wave in one scene?

The surface of the moon. (Or, as some believe, a secret film studio on Earth.)

On the moon, any such flag-waving would be totally impossible due to its vacuum environment. It was this that led a highly suspicious Kaysing to strongly suspect that the footage was not shot on the moon after all, but upon a world with a very demonstrable atmosphere—namely, Earth.

Unfortunately for both Kaysing and his dedicated band of followers and devotees, his claims became ever more sensational, controversial, and outrageous as the years passed by. For example, he later vocally asserted that NASA had been directly responsible for both the fatal January 21, 1967 fire on *Apollo 1* that killed astronauts Virgil Grissom, Edward White, and Roger B. Chaffee, and the equally fatal *Challenger* space shuttle disaster of January 28, 1986.

Kaysing's personal reasoning on these matters was deadly simple, and, in his mind, bolstered his initial theories concerning the Apollo Moon landings. The *Apollo 1* and *Challenger* astronauts, Kaysing came to conclude and accept, may very well have stumbled across undeniable evidence of NASA's secret special-effects trickery and, as a consequence, had to be silenced, no matter what the terrible costs to life, to limb, and to the future of the NASA space program.

CLASSIFIED

To bolster his inflammatory assertion that there were those within the corridors of power at NASA who would not think twice about killing Apollo astronauts to keep their secret, Kaysing noted, quite correctly, that after the *Apollo 1* fire, Thomas Ronald Baron, who had been a quality control and safety inspector for North American Aviation—the company that had built the Apollo command module—prepared a lengthy report on NASA safety protocol violation that was provided to a congressional committee investigating the fire.

Baron's report was followed by a further deadly tragedy for Baron and his entire family.

Authors Courtney G. Brooks, James M. Grimwood, and Loyd S. Swenson, Jr., said of this particular affair:

> Baron was a rank and file inspector at Kennedy from September 1965 until November 1966, when he asked for and received a leave of absence. He had made observations; had collected gossip, rumor, and critical comments from his fellow employees; and had written a set of condemnatory notes.[4]

Baron chose to follow a dual approach: First, he confided in his employers and work colleagues about what he perceived as clear and undeniable violations of safety-related issues on the part of NASA. Second, in Baron's mind, it was vital that the nation's media should come to understand the gravity of the situation. After spilling his guts to the press, however, the only thing that happened was that Baron was quickly fired from his job.

The team of Brooks, Grimwood, and Swenson further noted that:

> In the rebuttal, North American denied anything but partial validity to Baron's wide-ranging accusations, although some company officials later testified before Congress that about half of the charges were well-grounded. When the tragedy occurred, Baron was apparently in the process of expanding his 55-page paper into a 500-page report.[5]

On April 21, 1967, Baron had the opportunity to discuss his beliefs, concerns, and conclusions concerning NASA's safety issues with a congressional committee. It was all to no avail, however: Only one week after giving a deposition before the committee, Baron, his wife, and his stepdaughter were all killed when their car was violently hit by a high-speed train. The official verdict was that Baron, who had been behind the wheel of the vehicle at the time, had committed suicide, and his wife and stepdaughter were the tragic casualties of his overwhelming, personal selfishness and temporary loss of mind. For Bill Kaysing, however, this sorry and somewhat mysterious state of affairs was further evidence of a high-level conspiracy by NASA to hide the controversial truth behind the moon landings—as Kaysing believed, the total lack of moon landings, and also the lack of safety protocols, which would surely have had a major bearing on NASA's ability to get to the moon or not.

TOP SECRET

Years later, Kaysing was still provoking controversy, and maybe courting it too: in 1997, he initiated a legal action against NASA astronaut Jim Lovell, for libel, when Lovell described Kaysing and his attendant claims thus in 1996:

> The guy is wacky. His position makes me feel angry. We spent a lot of time getting ready to go to the moon. We spent a lot of money, we took great risks, and it's something everyone in this country should be proud of.[6]

Kaysing's case against Lovell was not destined to be successful, however; it was fully dismissed in 1999.

Kaysing passed away on April 21, 2005. At the time of his death, he was still championing his personal beliefs that the Apollo landings were complete and utter frauds.

CONFIDENTIAL

The 1978 release of the science-fiction/conspiracy movie *Capricorn One* added further fuel to the already scalding fire. In the film, the crew of *Capricorn One*—played onscreen by James Brolin, Sam Waterston, and O.J. Simpson—is awaiting the liftoff of the rocket that will take them on the first manned mission to the planet Mars. What the crew does not know, however, is that NASA has secretly learned that the ship's life-support system is destined to completely fail, and, as a result, sending the three astronauts to the red planet will be akin to signing their death warrant. The result: NASA is forced to fake the Martian landing from inside an old aircraft hangar situated at a decommissioned military base somewhere in Texas. At the time of this writing, a remake of *Capricorn One* is in the works, which is sure to fan the flames of controversy surrounding the Apollo moon landings.

CLASSIFIED

As most of the claims that NASA faked the moon landings were directly borne out of the research, theories, and observations of the late Bill Kaysing, the most important question of all is: How do they stand up to scrutiny? The answer is simple: They don't stand up to scrutiny. At all.

First, there is the matter of the Apollo astronauts' footprints, which at times looked to be very clear and even expertly carved as they walked, jumped, and paraded around the lunar surface, as is evidenced by some of the relevant photographs that NASA has now placed into the public domain. Conspiracy theorists suggest that, as a result of the moon's absence of water, the footprints should, at the very least, be vague in appearance. NASA argues that, not unlike wet sand, lunar soil is inclined to stick together. The result: perfect footprints. Or, as NASA might very well prefer to word it: Take that, conspiracy theorists.

Second, a great deal of hoopla has been made of the fact that that, in many of the pictures NASA insists were taken on the moon by its Apollo astronauts, the skies above show that the stars are not where they should be. Or, to put it more correctly, the stars are actually not there at all—none of them. The naysayers suggest that NASA would not have had the required skills, nevermind the capability, to ascertain the precise location of the stars on each and every one of the Apollo missions, so they took a much easier approach to resolving the vexing problem: They decided to completely leave the stars and planets out of all the relevant imagery.

NASA, however, has an answer to counter this particular allegation: When the photographs were taken, and the film footage was recorded, the overpowering brightness of the sun in an atmosphere-free environment, coupled with the daylight settings on the cameras, ensured that the stars overhead could not be seen.

The conspiracy theorists shake their heads in disbelief.

And what of the infamous waving flag that got Bill Kaysing so excited and enraged? There can be no denying that the American flag did seem to flutter after it was securely placed into the lunar surface by the *Apollo 11* astronauts—something that cannot occur within a vacuum. Although this issue, perhaps more than any other, has been seized upon by those who believe NASA deliberately engaged in massive fraud and deceit, the fact is that the flag ripples only for the very briefest of moments, and only when it is being handled by the astronauts themselves. Never again does it wave in the slightest, which is precisely what one would expect to see in a vacuum such as is present on the moon.

One of the key and critical issues that many conspiracy theorists forget, or simply have no awareness of, is that close to half a million people worked on the Apollo program. As NASA has quite reasonably pointed out time and again, the idea that any plan to fake the manned missions to the moon could have been successfully carried out, with not even a handful of those half a million personnel blowing the whistle on such a fantastic conspiracy is not just unlikely. Rather, it's nigh impossible. One only has to look at, for example, how quickly the Watergate, Iran-Contra, and Weapons of Mass Destruction fiascos collapsed and disintegrated. All of which involved far fewer players than those who worked on the Apollo program. For NASA to successfully buy the neverending silence of quite literally hundreds of thousands of people is simply not feasible. In fact, it's absurd. Moreover, even if only 2 percent of those employed on the Apollo program actually knew that the moon landings had been faked, we are still talking about close to 10,000 people. Again, could NASA convince all of those thousands of people not to ever talk? I consider the answer to be: most unlikely.

On this train of thought, whereas people such as Bill Kaysing have offered a variety of theories as to why they believe the moon landings were faked, the one thing that none of the conspiracy theorists have ever been able to successfully do is bring to the table, on the record, credible NASA sources who can provide complete, undeniable evidence of their knowledge and complicity in a program that resulted in the faking of each and every one of the Apollo moon landings—not even one.

TOP SECRET

The most persuasive data that favors the idea that NASA's Apollo astronauts really did set foot on the moon is the physical evidence that is presently at our disposal. Although the testimony of the astronauts, the photographs, and the film footage are not viewed positively or enthusiastically by many conspiracy theorists, far more difficult to dismiss is the large amount of lunar material that the astronauts collected on the moon and

A piece of moon rock brought back to Earth by the Apollo astronauts.

then brought back to Earth with them: namely, the samples of moon rock. Those are critical to the argument of whether or not humankind has ever successfully traveled to, and subsequently returned from, the surface of the moon.

Moon rocks are manifestly different in nature from any rocks that can be found on our home world, as Dr. David McKay of the Johnson Space Center, Texas, explained. He additionally said that the particular samples of rock that the Apollo astronauts brought back with them to Earth are dominated by very small craters that are clearly derived from meteoroid impacts of the type that can be found on the moon.

Of course, this information still fails to satisfy those who continue to argue that the human race is very much tethered to the Earth, and who believe that the moon landings were deliberately faked in some secret equivalent to a Hollywood studio. There is, however, still one more piece of evidence that seems to completely obliterate the views of the conspiracy theorists on this matter.

CONFIDENTIAL

On July 17, 2009, NASA unveiled photographs taken of the lunar surface by its Lunar Reconnaissance Orbiter spacecraft. Those photographs clearly showed evidence of the Apollo lunar module landing sites from decades earlier. Perhaps inevitably, however, the very fact that the Lunar Reconnaissance Orbiter is a project of NASA leads proponents of the "we never went to the moon" theory to outright dismiss the photographs as further highly skilled fakery.

CLASSIFIED

After careful study, the evidence does seem to be in NASA's favor, rather than those who cry "Foul!" But, to demonstrate that NASA is unlikely to ever convince everyone that it did indeed send its astronauts to the moon, take note of the following words of the former commander in chief and United States president, William Jefferson Clinton. In his autobiography of 2004, titled *My Life*, Clinton recorded that:

> ...Apollo 11 astronauts Buzz Aldrin and Neil Armstrong had...walked on the moon, beating by five months President Kennedy's goal of putting a man on the moon before the decade was out. [An] old carpenter asked me if I really believed it happened. I said sure, I saw it live on television. He disagreed; he said that he didn't believe it for a minute, that "them television fellers" could make things look real that weren't. Back then, I thought he was a crank. During my eight years in Washington, I saw some things on TV that made me wonder if he wasn't ahead of his time.[7]

It is no wonder that, as those words came from no less a source than a two-term U.S. president, NASA's assertions that it did send men to the moon are still met with deep suspicion and skepticism by millions.

TOP SECRET

There is one other aspect of the Apollo missions to the moon that was shrouded in deep secrecy for decades, but it is an aspect of the project that actually offers much support to the notion that the astronauts really did set foot on the surface of the moon: Although the Armstrong-Aldrin-Collins flight proved to be a spectacular and historic success, behind the scenes, both NASA and the White House knew that this never-before-attempted mission was an extremely dicey one. There was a very real possibility that it could all have ended in awful tragedy, with Armstrong and Aldrin stranded on the surface of the moon, faced with an inevitable and rapid death, while Collins, orbiting above the lunar surface, could have been forced to make the terrible decision to leave his friends and comrades behind and head back to the safety of the Earth, utterly alone. A brief speech was secretly drawn up for then-President Richard M. Nixon, to be given in the event that he was forced to reveal to the world that the worst-case scenario had tragically come to pass. A July 18, 1969 document sent from Nixon aide Bill Safire to Nixon's White House Chief of Staff, H.R. "Bob" Halderman, reveals the text of this secret speech. Given the title of "In Event of Moon Disaster," it states:

> Fate has ordained that the men who went to the moon to explore in peace will stay on the moon to rest in peace. These brave men, Neil Armstrong and Edwin Aldrin, know that there is no hope for their recovery. But they also know that there is hope for mankind in their sacrifice. These two men are laying down their lives in mankind's most noble goal: the search for truth and understanding. They will be mourned by their families and friends; they will be mourned by the people of the world; they will be mourned by a Mother Earth that dared send two of her sons into the unknown. In their exploration, they steered the people of the world to feel as one; in their sacrifice, they bind more tightly the brotherhood of man. In ancient days, men looked at stars and saw their heroes in the constellations. In modern times, we do much the same, but our heroes are epic men of flesh and blood. Others will follow and surely find their way home. Man's search will not be denied. But these men were the first, and they will remain the foremost in our hearts. For every human being who looks up at the moon in the nights to come will know that there is some corner of the world that is forever mankind.[8]

Thankfully, the speech was never needed, and the secret documents languished in obscurity, all but forgotten for decades.

Chapter 5
The Area 51 Connection

The following, startling account of John, born in Bloomington, Minnesota, a man who served from 1948 to 1959 in the New York Police Department, has a major bearing upon the work of NASA in relation to ultra-secret data on UFOs and alien beings. His account also offers intriguing data suggesting a connection between NASA and secret, UFO-dominated projects undertaken at the infamous Area 51 installation in Nevada.

I first encountered John in the latter part of 2005, after he read my book *Body Snatchers in the Desert*. The book was a study of the notorious UFO crash at Roswell, New Mexico, in the summer of 1947. *Body Snatchers*, however, took a distinctly different approach to its subject matter, and was focused upon allegations that the Roswell crash had far less to do with aliens and much more to do with secret, high-altitude balloon experiments using Japanese prisoners of war and physically handicapped people. When John contacted me, he explained that during the course of his work with the U.S. intelligence community in the early 1970s, he came across certain data that seemed to dovetail with some of the material contained in my book. Concerned by the possibility that if he spoke on the record he might very well lose his pension from his former employer, Wackenhut, John agreed to a somewhat off-the-record interview, which was conducted in 2006, and reveals the many twists and turns of his story.

John stated that from the latter part of 1957 until the early months of 1958 he and a number of police colleagues were tangentially involved in an FBI undercover operation to try and locate and arrest a Soviet spy ring that was believed to be targeting and infiltrating a U.S. defense-related company in the city. It was suspected that someone within the company had been compromised by the Soviets, although this was never ultimately proven. The police were involved because the compromised individual also had suspected links (which, again, were never ultimately proven) with a money-laundering Mob-based operation.

John explained that arrests of several Soviet citizens were made in connection with this particular operation, along with at least two American citizens. Partly as a result of his involvement in this operation, and also because he was possessed of a very ambitious character, after leaving the New York Police, John was involved in additional work with the FBI that began in late 1959 and continued, albeit sporadically, until October 1970. John declined to describe fully the nature of his work with the FBI, but stated that it centered upon Soviet attempts to obtain classified files of an intelligence, military, and defense nature from within the borders of the United States.

In July of 1970, John was approached by several of his old colleagues, and was duly offered a very lucrative position within Wackenhut, a company that came into being in 1954 and, to this day (albeit now under the control of the Danish organization Group 4 Falck, which obtained Wackenhut in 2002) offers security-based services to major private and governmental bodies. Indeed, as far back as 1964, Wackenhut was providing welcome security at NASA's Florida-based Kennedy Space Center. John began working with Wackenhut in December 1970, and in February 1971 was asked if he would be interested in doing contract work for a certain highly secret arm of the U.S. intelligence community operating out of Nevada.

John was advised that the salary would be very good indeed, that the contract would be for one year only, that the work would ensure for him very significant contacts within the intelligence community, and that it would position him very well for future career prospects. After speaking with his bosses at Wackenhut to secure their opinions and thoughts on the matter, who all strongly urged him to accept the job, John did precisely that.

John said that, on accepting the new position, he was very soon subjected to a stringent background check, and was interviewed on six occasions—twice by NASA security personnel, twice by the representatives of an unnamed agency, once by someone who identified himself as working for the National Security Agency (NSA), and once by a former senior colleague at the FBI. The results of the interviews were apparently satisfactory, and John began working, in April 1971, at the location in Nevada that he later came to realize was none other than what, today, is infamously referred to as Area 51.

John revealed that a part of the reason he was given the posting was that, as a bachelor, he could remain on base for 27-day stints at a time as required, and could then return to Las Vegas for only three or four days—depending on whether there were 30 or 31 days in the month—before returning to the secret location. This, John said, would not have been the ideal situation for a married man, or someone with children. Notably, John was willing to admit to me that those involved in the security screening process knew that while back in Las Vegas John frequented high-priced prostitutes, but this was not seen as problematic to his posting at all. Rather, the fact that John had no family ties whatsoever was perceived as a bonus when it came to maintaining security out at Area 51. As John said,

> Even if I'd said something to a hooker about aliens and UFOs, which I wouldn't have done anyway, who was going to believe them?[1]

John was always flown both into and out of Area 51 in a small aircraft, usually a Cessna, along with three colleagues whom he got to know very well in the one year they worked together. There were blinds on the windows of the aircraft that had to be pulled down after takeoff from Las Vegas and that had to remain pulled down after landing too. And, upon landing at Area 51, everyone was required to put on a pair of goggles. These were not normal goggles, however. In fact, they were quite unlike any that John, and his newly found friends and colleagues, had ever seen before.

The goggles had split lenses, not unlike bifocal glasses, and the top section was so thick that it prevented any effective vision, and all that one could make out was a vague blur. The only part of the goggles that could be seen through with any real effectiveness was a small section at the bottom of each lens. As an inevitable result, the wearer was forced to look only in the direction of the ground, and their shoes, to see where they were going. This was without doubt done deliberately, John stated, to ensure that the person wearing the goggles was unable to make out the particular details of anything that he might inadvertently view on the base, and which he did not have official clearance to see. The group was always guided to a bus, again with blacked-out windows, and was then driven for a few minutes to yet another location. When getting off the bus they were still required to keep the goggles on and were directed to their place of work.

The location in question was a small, concrete building approximately 60 feet square. Security at the door was stringent, John added. Upon entering the building, which was simply a bare concrete room, he was finally allowed to take off his goggles. He viewed two things only: the entrance to an elevator, and a concrete staircase that only descended. Sometimes, John said, they would enter the elevator, and sometimes they would use the staircase to descend the two floors to their place of work. He had no idea at all what work was carried out on the first floor below the surface. On arriving at the second floor, there was a second, rigorous security screening procedure—a procedure that would follow every time he returned to the base after his three to four days back in Las Vegas. He recalled that after passing through the second security screening, he entered, via a large pair of thick, metal doors, a long corridor that was around 180 feet in length, and that had six doors leading off it, three on the left and three on the right. His office was the second on the right, and it had a simple lock, identical to a standard lock on the front door of a typical house.

John was directed to his new office, and was given an initial briefing by three men, all displaying official NASA credentials. He was advised that the location in which he would be employed was officially classed as the History Department. The three NASA men relayed to John that the overall location at which he was working (he says that the term *Area 51* was never used during his employment there; it was only years later that the term became widely used and known to him) was involved in the development of prototype and radical aircraft, biological warfare agents, chemical warfare technologies, exotic weaponry, and "something else."[2]

That "something else," John was told, was research into "things that had happened in the 1940s." It very soon became clear to John and his colleagues that they were being employed as custodians of historical files on these subjects—files that would, from time to time, be required by base employees during the nature of their work. John stated that there were three reasons, and three reasons only, that the files of which he was the custodian were ever accessed by anyone:

1. On a regular basis scientific personnel would require access to the files as part of their work.
2. The files would be used to brief new employees to the base if such things were deemed necessary.
3. There would be regular, random checks on the part of security personnel to ensure that all of the files were present and had not been tampered with in any fashion.[3]

According to John, the priceless documentation was never allowed out of his office, and the briefings for new personnel were always conducted inside the confines of his office by a four-person team from NASA. And if anyone ever needed to access the files as part of their work, they would have to make pencil notes from the files, and pencil notes only. Those pencil notes had to be exclusively written on a bright, orange-colored paper of an unusually large size, a type John had never seen before or since. He speculated that perhaps this curious color and size could somehow have been picked up on security screenings if anyone tried to smuggle the paper out of the base.

John stated that one of his colleagues became the custodian of historical files on prototype aircraft at the base, another had jurisdiction over historical files on exotic weaponry being developed there, and another was responsible for historical files on biological weaponry. John's role was to maintain the historical files on UFOs. He says that the first two weeks were spent receiving extensive briefings from personnel on the base, and, due to the fact that he was the custodian of such material, he was allowed to read all of the files, with absolutely no restrictions at all, and acquaint himself with the document names and numbers in a fashion that would allow him to identify them with ease if they were ever requested by visitors to his office.

John said that the files covered the period from 1943 to 1968. He added that, as did most people in that era who worked in the official world, he knew that the U.S. Air Force had investigated UFO sightings via its Project Blue Book study that finally closed its doors in 1969. He elaborated that the files in his office told a far more in-depth and deeply sinister story of official, and highly classified UFO investigations in the United States that went far beyond anything and everything in which Project Blue Book had even been involved. John stressed several times that he never saw any UFOs, any live aliens, or any alien bodies at Area 51—only documents that dealt with such matters from the specific time period of the 1940s onwards. However, the fact that there were people who required access to the files led him to believe that there were indeed crashed UFOs and alien bodies on the base, unless this was all evidence of a gigantic and ingenious ruse to test his loyalty, which, he admitted to me, had crossed his mind on more than several occasions.

According to what John learned during his time spent at the base digging deep into the secret documentation, in 1947, an incredibly well-hidden group was established deep within the confines of the American intelligence community, which, from 1960 onward, worked closely with NASA on matters of a presumed extraterrestrial nature. In the late 1960s (John was unsure whether it was in 1968 or 1969) the group was radically reorganized. Instead of a number of bodies, such as NASA and the CIA, playing integral parts in the program, but retaining all of their own files and their own input in the process, all of the work, data, and documents were subsequently transferred to a centralized body that operated out of what is now known as Area 51. And it was also at Area 51 that certain NASA, CIA, NSA, and Air Force personnel were offered full-time positions that would take them far away from their normal day-to-day activities in their relevant agencies.

Yes, said John, the new group continued to have the same input from intelligence personnel that it had before, and much of the membership stayed exactly the same. The only major difference was that all of the documentation was removed from the vaults of the different intelligence agencies, and from NASA, and was all taken to Area 51 for consolidation. Anyone in NASA, in the intelligence field, and in the military who worked on the projects was required to transfer to Area 51 too.

John felt strongly that it is for this chief reason that no one, to date, has ever been able to access any data that conclusively proves NASA has been engaged in high-level UFO-connected conspiracies: All of the relevant records that might have been related to such issues, he fully believed, were removed from NASA's vaults back in the late 1960s, and, as a consequence, will never, ever surface for official declassification under the Freedom of Information Act legislation.

Certain people—very influential and power-wielding people—allied to the original program were considered to have gone rogue to a significant and deadly degree, and had gone way beyond their normal remit to keep certain UFO secrets hidden from the public, the press, and even from elected government officials (which may have included the office of the president of the United States). This, said John, was part of the reason for having a centralized location for the many files and documents: A number of them referenced a variety of controversial and highly secret actions that had been undertaken to hide the alien truth. These were, in many ways, more sensational than the UFOs and their alien crews themselves. If these files were scattered around the vaults of numerous agencies, explained John, there was a very strong chance that at least some of them might eventually leak outside of official circles. Those rogue individuals, he revealed, had not complied with orders to relinquish their files, and, as a result, some files were found to be missing from secure vaults. Other documentation, supposedly, had been shredded, but it was strongly suspected that they had not been destroyed after all. On top of that, certain people with clearance to see such material had vanished, never to be seen again, or were found dead under questionable circumstances.

John recalled that the files he read all had one curious similarity: The cover page on each and every one was missing. He explained that all of the files were stapled, and many of the documents had a tiny piece of torn paper stuck under the staples, clearly indicating that the cover pages had been carelessly torn off. In turn, a new cover page had been added—a blank piece of paper with a new pencil-written title affixed with a paperclip to the second page.

He was told that these copies had come from the office of a man who had a major role to play in secret UFO investigations in the late 1950s and early 1960s, and, in addition to the files he had legal access to, the man had obtained some of them after they had been stolen from the vaults of other U.S. intelligence and military-based agencies. He then kept them in a secure location and assigned his own filing and naming system to the files. The man was also suspected of having shredded numerous documents that all dated from the mid 1940s for private, obscure reasons of his own. These files were never duplicated, and told stories that are now, one is almost forced to presume, utterly lost forever. John alluded to the possibility that that this man was none other than the infamous James Jesus Angleton, who was appointed Counter Intelligence Chief of the Central Intelligence Agency by CIA Director Allen Dulles in 1954.

CONFIDENTIAL

The earliest document John recalled was, he said, without doubt the most disturbing one too. It dated from 1944 and was essentially a very long letter of complaint to the wartime Office of Strategic Services (OSS) from a certain medical man who was then working at the secret Los Alamos Laboratory in New Mexico, and whose team had been required to carry out tests and experiments on a group of very unusual-looking people who had been secretly brought to the installation. In the letter, the doctor was complaining to a military officer about the nature of what was being asked of his team, and was extremely concerned about why these people looked the way they did.

John said he remembered clearly reading that, at some point in 1943, 17 people of highly strange appearance were brought to Los Alamos. They all looked very much alike: around 5 feet in height, hairless, with enlarged heads and oversized eyes. They were brought to the base in a covered Army vehicle, were totally naked, and walked in a stiff, jerky fashion. He stressed that the eyes of the people were normal apart from their size—in other words, they were not the large, black, hypnotic eyes that are usually ascribed to the stereotypical diminutive alien that dominates today's popular culture. The group of people, or creatures, was taken to a secure location at Los Alamos, and throughout the course of four months was subjected to a variety of medical tests. Blood was extracted, skin samples were taken, and several of the beings were used in radiation experiments and bacteriological warfare tests—a controversial new field of science.

The doctor described all of this in a truly nightmarish fashion: the people never, ever spoke, and only made strange noises like a cross between a seal's bark and a hiccup that was very loud and highly disturbing. They would have to be restrained by military personnel before any of the tests could be performed, and they were fed on a diet of a mashed and condensed mixture of various fruits, and occasionally mashed potato, and drank only water and milk. According to the doctor, he was told by the military that they were malformed people who had been secretly taken from asylums and hospitals for use in classified military experimentation, and nothing else.

As a medical man, the doctor wrote that he found the explanation to be highly suspicious. There was, he believed, very little likelihood that 17 people could all be afflicted in such an extremely strange fashion, and the world's mainstream medical community didn't know anything of it. The doctor advised in the pages of his letter that rumors at Los Alamos suggested that no one—not even the military who brought them to Los Alamos—knew who or what these people were or where they were from, but they had apparently been found wandering around somewhere in the Arizona desert. All of the 17 people died before the end of 1943, and their bodies were carefully removed for autopsy by the military as they died, one by one. No one was ever able to communicate with any of the 17 people, and the doctor complained bitterly about the way in which he and his team were threatened, physically and mentally, by the military and told never to discuss the matter of the strange beings outside of the Los Alamos installation.

The file, said John, referred to additional documents on these strange-looking people that, he was told in his briefing, had been destroyed by the man from whom they had originally been taken. Bizarrely, John was informed, no record of these diabolical experiments now existed, apart from this one solitary letter. In addition, the location of the bodies of the 17 creatures, or people, was now unknown, and the entire paper trail appeared to have been destroyed decades ago by the people who had been involved and implicated in the project. It was suspected, however—albeit only by word of mouth and institutional memory—that, in some fashion, the beings found in the Arizona desert in 1943 were directly linked to the dead bodies of unusual appearance that were said to have been recovered by the U.S. military near Roswell, New Mexico, in the summer of 1947.

TOP SECRET

There was also a very large file at John's disposal titled "Autopsies—Bodies Unknown Origin 47" that dealt with the bodies of eight equally strange-looking people found in the New Mexico desert (there was no reference in the autopsy report to an accompanying spacecraft or a vehicle at all) in the summer of 1947, which John had the opportunity to read at length. All of the "people" were substantially like the creatures taken to Los Alamos in 1943. There was no name on the document to identify the writer; however, John suspects that the name was probably on the missing original title page. He also noted, with some significance, that a small NASA crest appeared on each and every page. Given the fact that, as internal references made very clear, the documentation had been generated some years before NASA had even come into being, the aforementioned crests had to have been added at a later date—perhaps, speculated John, when NASA managed to obtain a copy of the document for its own scrutiny and evaluation.

John also recalled that the file stated that of the eight bodies, four were preserved pretty much intact; two were autopsied, and their remains were then completely disposed of; and two were dissected, and their body parts and organs were carefully cataloged and preserved separately. John said that the document referred to a series of both color and black-and-white photographs of the beings, of the autopsies, and of their organs; however, at no time did he ever see any photographs in any of the files in his possession. Several of the bodies were extremely badly damaged—in a fashion that suggested some form of violent accident had occurred—and the rest were in relatively, and surprisingly, good condition, it was claimed. The report was very detailed and covered the major bodily organs: the brain, the eyes, the ears, the skin (which was described as being a sickly, white-gray color), as well as the fingers and toes, blood analysis, the limbs, and the teeth of the entities.

CLASSIFIED

Interestingly, there was another top-secret file that had the handwritten title of "Autopsies—Bodies Unknown Origin 47, Biological Problems and Deaths" that dealt with an unknown virus that had reportedly contaminated and killed an entire medical team that had examined one of the bodies. All of the medical personnel had the foresight to wear protective suits, but these were apparently completely useless when it came to providing adequate biological protection. Curiously, none of the other bodies exhibited any signs of this unidentified, killer virus.

CONFIDENTIAL

John stated that one report in particular interested him greatly. It was titled "Suit Study 48 Armageddon" and was a lengthy document that was focused upon a scientific study of the clothing that had been found on certain bodies recovered in New Mexico in 1945. That same clothing was a yellowish color and was a one-piece outfit that extended from the lower neck right down to the bottom of the feet. It apparently took several hours to remove the clothing of the creatures by forcibly cutting through the material. Upon examination, it was determined that the suits were put together in what today we would term a Velcro-style fashion. However, the suits almost seemed to be alive or had an in-built memory, because, not only would the individual fibers bond when brought close together, but they also appeared to bond with the exact same corresponding fibers time and time again.

The most bizarre aspect of this file centered around one of the scientific personnel—the shortest member of the team—who chose to try one of the suits on. Having finally learned how to undo the suit with ease, his colleagues held it open for him to climb into. Somewhat alarmingly, the suit quickly molded itself around the man. Due to his size, the suit was not comfortable but could at least be worn. Something very curious and ominous occurred when the outfit was donned, however. The man began to feel very claustrophobic and started to receive disturbing images in his mind of a dark and frightening future for the Earth and for all life on the planet, and particularly so for the human race. It was a future that was dominated by an irradiated world, ruined cities, huge atomic mushroom clouds looming miles upward into an ever-black sky, as strange objects resembling flying saucers flew across the ravaged landscape. Perhaps most worryingly of all, the human race had been reduced to absolutely minimal levels as a direct result of a deadly virus of unknown origins that targeted and destroyed the human immune system. The man in the suit received a distinct and disturbing impression that whatever these creatures were, they hated us to a truly profound degree, and there was some sort of plan in the works on their part to try and spark off a nuclear holocaust between the United States and the Soviets, as a means to "take us all out [of] the picture."[4]

Correctly sensing that he was in deep distress, the man's colleagues and friends quickly ripped the suit off him. A report was soon prepared, noting carefully that the man was briefed intensively by a team comprising

NASA and CIA personnel with particular expertise in matters of a psychological nature. The file also stated that this bizarre episode was a further clue to the fact that the suit was somehow alive, and appeared to be linked with the minds and memories of the creatures that had previously worn them. Perhaps most weirdest of all was a note in the file that stated the suits were thereafter kept locked away in a secure, guarded vault, and were considered to be nothing less than "intelligent hazards."[5]

TOP SECRET

Most of the additional files, recalled John, fell into two particular categories:

1. There were dozens and dozens of very technical papers on two glider-type aircraft that had been found in and retrieved from the New Mexico desert in 1947. These dealt with structure, landing gear, and much more that John couldn't recall.
2. There were also literally hundreds of one- and two-page memos from and to various personnel discussing ongoing projects, many of which focused upon biological and bacteriological issues and UFOs.

John said that there was never any reference in the files to the creatures or the aircraft being definitively extraterrestrial in origin. Indeed, no one seemed to have any solid awareness of what they were, or what might have been their point of origin.

CLASSIFIED

John told me that there was a small collection of documents dating from July 1947 speculating that this might have all been the result of an ingenious hoax on the part of the Soviets—until, that is, it very quickly became apparent to one and all that not even the Soviet Union would have had the required expertise to successfully pull off such a fantastic ruse, much less biologically alter, or mutate, a number of human beings into something so different. Another theory mentioned in the records, John asserted, was that this was all part of a U.S. cover story to hide far more down-to-earth Cold War atrocities undertaken on mentally and physically handicapped people.

John stated that among the people with whom he worked, however, there were three possibilities that were considered as being among the most interesting:

1. These things were from another world entirely.
2. They were from somewhere on the Earth—other dimensions or the more remote parts of the world were discussed.
3. They were, incredibly, from our very own distant future.

John added that, in discussion with colleagues at Area 51, he learned that even by the turn of the 1970s there was still no definitive proof as to what these creatures were or from where they came. But there was one point that, for some of the personnel, strengthened the time-travel idea: Although the glider aircraft recovered in New Mexico were most certainly highly advanced, technologically speaking, several components within the craft were constructed in feet and inch measurements—a system that, John concluded, aliens from a distant planet would simply not employ. In other words, the craft themselves, and the entities that built them, seemed to be connected to us to a significant degree.

Also, there were references in the files to extensive studies done on the possible existence of alien life, a subject upon which numerous NASA experts had reportedly been consulted. The overriding conclusions of the NASA specialists were that there was nothing to deny the possible existence of extraterrestrial life in the Universe, that it was even possible that such life was coming here, and that this is precisely what these creatures were, after all.

The most suspicious aspect of this whole affair for NASA, however, was that not only were these creatures apparently completely comfortable breathing our air, but they were also perfectly synchronized for the Earth's atmospheric pressure too—which seemed to be too coincidental. John said that this information decisively split the teams into two camps—one that favored the idea that the creatures were from Earth, either from a time far in the future or from some secret location on the Earth in our present, and one that believed that they were indeed aliens that had been genetically altered, in some truly radical fashion that involved a science far beyond our own, to allow them to comfortably operate on our planet.

John remained with the program for the prearranged one-year contract and then carved out a career for himself in the private sector (which included doing security background checks on people applying for jobs with NASA) before finally retiring in the winter of 1981. His thoughts and theories on this curious affair, as well as his exposure to the files at issue, are intriguing. As he pointed out, he had absolutely no expertise in the area of UFOs whatsoever. And so, why, if this was such a huge and important secret, was he even briefed on the subject matter in the first place? In addition, he wondered: Given the alleged magnitude and significance of the operation, is this not the type of job that a person would have been drafted into for his entire working life, rather than just for a mere 12 months at most?

Taking these facts into careful consideration, John stressed that although the documentation at issue certainly looked genuine, he was never able to entirely dismiss from his mind the possibility that his exposure to the files could have been a part of some large, curious, and convoluted mind game on the part of NASA and the intelligence services such as the CIA, Air Force Intelligence, and the NSA. Because his work at Area 51 and his access to the files came about as a direct result of his FBI contacts, John speculated that his superiors may have exposed him to totally bogus materials at Area 51, and then watched his every move to see if he spoke out of turn. The fact that John never did speak out of turn in that 12-month period and was thereafter considered utterly trustworthy led him to be rewarded with a near decade-long career in the private security sector. It was a career that saw him move, practically effortlessly, within highly influential circles in the world of U.S. intelligence that were totally unconnected to UFOs.

The truth may really be out there, but it appears to be a truth that is guarded and protected by an infinite number of hall-of-mirrors-style cover stories.

Chapter 6
An Alien Epidemic

During the latter part of April 2003, while I was undertaking research into specific rumors that NASA had undertaken classified research into potentially lethal alien viruses, I posted a carefully worded message to various members of an online discussion group that I was then subscribed to. My message focused on biological and chemical warfare. In the wake of this action, I received a telephone call from a man named Ralph Jameson, who, several decades ago, had been employed at Fort Detrick, Maryland, home to the U.S. government's defensive research into the field of biological warfare.

Jameson told me that the 1969 book *The Andromeda Strain*, written by the late Michael Crichton, and the subsequent 1971 movie spinoff of the same name, were of deep interest and concern to staff at Fort Detrick because of the central theme of the book: namely a terrestrial space vehicle returning to Earth containing deadly alien microbes that subsequently unleash a lethal alien virus upon the human population.

Jameson openly claimed to me that approximately nine months after the movie version of *The Andromeda Strain* was released, a document came across his desk titled *Andromeda: Fact or Fiction?* Jameson was keen to stress that the document was not an officially sanctioned document. Instead it had been wholly researched and prepared by an employee of Fort Detrick in his free time. The man had, however, reportedly submitted the document to his superiors in the event that it might one day be deemed of interest to Fort Detrick employees, as well as for future research and consideration. Jameson further elaborated that the report addressed *The Andromeda Strain* scenario very closely, and looked deeply into the matter of whether or not such a nightmare might one day become a terrifying reality. The report ran to just under 200 pages, and although it did not discuss UFOs in any way, shape, or form, it did concentrate on two controversial issues:

1. An alien virus finding its way to Earth, possibly on a meteorite.
2. The total contamination of the Earth and all its many and varied life-forms by an alien virus as a direct result of NASA's astronauts bringing back samples from the moon, or, perhaps later, from the planet Mars.

The report allegedly dug deeply into:

- ✓ The way in which an out-of-control extraterrestrial virus might spread wildly.
- ✓ How world authorities might be forced to deal with it.
- ✓ What precautions would need to be taken if matters escalated.
- ✓ How a viable vaccine might be created, if at all possible.

The report also discussed how just such a possible vaccine might remain elusive to us, the resultant effect being that the virus could very possibly mutate into a form of a doomsday virus for the entire human race. This unsettling report, Jameson explained, was placed into the Fort Detrick document archive and was occasionally pulled out and read by interested parties that were employed there.

Jameson additionally stated that although the original copy of the document never left Fort Detrick, on one occasion something distinctly strange occurred: One of the three copies of the document that had been made at the time was forwarded first to NASA and then to the CIA's Office of Science and Technology, and remained absent for several months. Interestingly, when the document was returned to Fort Detrick, it quickly became clear that its contents had been closely scrutinized, to the extent that various red-pencil changes had been made to the document, and certain sections had been carefully underlined by both the NASA and CIA personnel that had requested access to the material.

One person who apparently had full knowledge of the history of the document, as well as its content and implications, said Jameson, was Dr. Charles Rush Phillips, of whom, Fort Detrick authority Norman Covert stated:

> His work between 1943 and 1969 made it possible for scientists to have and maintain the tools they needed to develop medical knowledge about microorganisms. The result has been the development of vaccines for a variety of diseases and an understanding of how disease spreads.[1]

TOP SECRET

As fantastic as it might sound for NASA, the CIA, and Fort Detrick personnel to express deep interest in and profound concern about alien viruses, it is actually not overly unusual. In fact, extraterrestrial disease is a subject that has worried some of the finest minds within the scientific world for years.

According to Article IX of the "Treaty on Principles Governing the Activities of States in the Exploration and Use of Outer Space, Including the Moon and Other Celestial Bodies," which was collectively signed at Washington, D.C., London, and Moscow on January 27, 1967, and which was entered into force on October 10 of that same year:

> In the exploration and use of outer space, including the Moon and other celestial bodies, States Parties to the Treaty shall be guided by the principle of co-operation and mutual assistance and shall conduct all their activities in outer space, including the Moon and other celestial bodies, with due regard to the corresponding interests of all other States Parties to the Treaty. States Parties to the Treaty shall pursue studies of outer space, including the Moon and other celestial bodies, and conduct exploration of them so as to avoid their harmful contamination and also adverse changes

> in the environment of the Earth resulting from the introduction of extraterrestrial matter and, where necessary, shall adopt appropriate measures for this purpose.

The deep concern expressed here about "adverse changes in the environment of the Earth resulting from the introduction of extraterrestrial matter" was, and still is, very real, and greatly warranted.

CLASSIFIED

One of those who recognized, decades ago even, the potential, major threat posed to the human race by outbreaks of lethal viruses of extraterrestrial origins was Joshua Lederberg, who received the Nobel Peace Prize for his studies, and who worked on NASA's experimental programs to ambitiously seek out life on Mars. An eight-page paper written by Lederberg for *Science* on August 12, 1960, titled "Exobiology: Approaches to Life Beyond the Earth" bleakly stated that: "At present the prospects for treating a returning vehicle to neutralize any possible hazard are at best marginal by comparison with the immensity of the risk."

Then, on July 10, 1969, Lederberg wrote a letter to *The New York Times* (published on the 13th) and castigated the newspaper for discussing the possibility that the NASA astronauts that returned from the *Apollo 11* moon-landing mission, Armstrong, Aldrin, and Collins, offered "a tangible risk of global infection by lunar microbes." Lederberg carefully explained in his letter that the complete absence of an atmosphere effectively rendered it almost impossible for the moon to harbor deadly microbes or viruses that could have an effect on life, human or otherwise, on our planet. He was very careful to stress, however, that the issue of having to "protect the Earth against possible infection from Mars," was one that definitely needed to be carefully studied and resolved, and that required far more extensive and rigorous study if NASA should one day decide to send astronauts to explore the red planet.[2]

Published in the *Washington Post* on August 16, 1970, a two-page feature from Lederberg titled "Engineering Viruses for Health or Warfare" addressed the issue of modified and deliberately created viruses for use in warfare, and similar modifications for health purposes. Lederberg stated in part:

> We now begin to realize that the intentional release of an infectious particle, be it a virus or bacterium, from the confines of the laboratory or of medical practice must be formally condemned as an irresponsible threat against the whole human community.[3]

On this matter Lederberg was far from being a solitary voice. Writer Leslie Mullen, for example, noted in 2003 that:

> Even though there is no proof of bacterial or viral pathogens anywhere except Earth, there is already a worried advocacy group called the "International Committee Against Martian Sample Return," and science fiction novels like The Andromeda Strain depict nightmare alien infection scenarios.[4]

Similarly, in June 2003, *National Geographic News* revealed that in a letter sent to the British medical journal *The Lancet*, Chandra Wickramasinghe, of Wales's Cardiff University, offered the disturbing theory that the SARS outbreak might very well have originated in outer space, and then made its way to Earth, before coming down in China, which is where the outbreak originally kicked off.[5]

In view of these cautions, we would be wise indeed to take careful note and heed of the following words from "Medical Science, Infectious Disease and the Unity of Humankind," a two-page article that Joshua Lederberg penned for *The Journal of the American Medical Association* on August 5, 1988, which dealt with his ever-present concerns about super-viruses wreaking havoc across the globe. Lederberg noted, in the final paragraph of this particular paper that:

> The microbe that felled one child in a distant continent yesterday can reach yours today and seed a global pandemic tomorrow.

We have been duly warned. Whether we are fully prepared for any and all such possibilities, however, is an entirely different matter altogether.

Chapter 7
1970s Europe

Firm evidence of the fact that NASA personnel, during the early to mid 1970s, were deeply interested in seemingly bona fide UFO encounters can be found within the pages of a formerly classified Department of Defense intelligence report dated August 22, 1974, which was shared with senior NASA personnel at the agency's headquarters. The document in question details a wealth of truly extraordinary UFO data emanating from the country of Spain in a clearly delineated time period; namely, the fall of 1973 to the early summer of 1974.

The documentation forwarded to NASA begins as follows:

> During the period September 1973 to June 1974 a rash of UFOs appeared over Spain and sightings were reported by various types of people.[1]

And with that said, let us focus on those UFO case files from Spain that NASA deemed to be so important to its understanding of the complexities of the UFO phenomenon.

CLASSIFIED

The first report was a highly curious one indeed, and concerned the sighting by several witnesses late one September night in the El Ferrol del Caudillo region of Spain of an approximately 3-foot diameter, brightly lit, and circular-shaped UFO that appeared to be maneuvering precariously at near ground level on two leg-like appendages. Was this object, taking into consideration its very small size, some form of remotely piloted alien device, not unlike the type that are now, nearly 40 years later, routinely deployed in and around the skies of Iraq and Afghanistan by our very own military? Was it, perhaps, not unlike the very small UFO that was viewed by NASA astronaut "Deke" Slayton in the early 1950s? Whatever the answers to that question may be, the UFO was not intent on hanging around for long, and suddenly took to the skies and vanished into the darkness.

Spain was not destined to be free of UFOs for very long, however, learned NASA. Three months later, early on the morning of December 11, 1973, a classic flying saucer–style vehicle was seen at Malaga, Spain. It hovered in the dawn skies for approximately a quarter of an hour, and then, without warning, suddenly shot away at high speed in the direction of Torremolinos. But, it was in March and April of 1974 that matters really began to heat up on the UFO front, and fully ensured that NASA sat up and took notice of what was afoot on the other side of the world.

After a few relatively insubstantial reports of vague, unusual aerial lights surfaced in the first two weeks of March 1974, things proceeded to change drastically and in a highly unanticipated fashion. Certainly, one of the most amazing of all cases (the details of which were shared with NASA) was that which took place on March 23, 1974, some three miles to the south of Castille de las Guardas.

Incredibly, the witnesses, which included two police officers and a car salesman, saw what were described in the official files as a trio of gigantic, aluminum-colored craft, broadly described as being the shape of huge pencils, 500 to 600 feet in length, and flying very slowly and at near perilously low levels. Suddenly, three much smaller craft, which all the witnesses agreed resembled mushrooms, flew out of one of the huge craft. In complete and utter silence, they proceeded to practically plummet to the ground, then rose slightly, and began to pursue the salesman, who, in terror, fled the scene on foot. Fortunately, no harm was done to the man, and the mushroom-shaped craft all vanished skyward, as he finally reached the outskirts of the village of Castillo de las Gaurdas.

All this was an astonishing taste of what was soon to come—very soon, in fact; it was only hours later that the next extraordinary encounter to pique NASA's interest took place.

This time, the witness was the official chauffeur of the president of the Cadis Provincial Commission. Darkness had fallen on a particular stretch of highway located near Sanlucar de Barrameda when the chauffeur witnessed a brightly lit object dancing wildly around the skies at a height estimated to be barely 50 feet above the road. Not only that: the UFO nearly disabled the entire electrical system and engine of the man's car as he drove slowly and cautiously in its immediate direction. Notably, this seeming ability of the UFO, or perhaps its alien crew, to adversely affect the man's vehicle closely paralleled a statement the FBI had made way back in 1957, in the wake of the Soviet Union's launch of *Sputnik 1*. As detailed in Chapter 1 of this book, at the time of the Russian launch, the FBI noted that:

> ...persons in the Southwestern states while driving their cars have allegedly seen UFOs that caused the engines in their automobiles to stop.[2]

For the next few days, the skies of Spain were filled to the brim with unknown aerial objects, with hovering balls of light, and with small, illuminated, triangular-shaped machines that seemed to rock slowly in the air like boats gently bobbing on the waves of a tranquil lake. Things weren't destined to stay tranquil for very long, though. It was soon time for the aliens themselves to put in an appearance.

TOP SECRET

The night of March 26, 1974, was when all hell broke loose. In this case, it was a young truck driver barely out of his teens who had the distinct misfortune to encounter the UFO phenomenon in all its strange and unearthly glory. As he headed home along the highway at Valdehijaderos, a circular-shaped object—described in the documents forwarded to NASA as looking like two plates placed together—that was clearly under some form of intelligent control, appeared in the sky at an extremely low altitude. In fact, so low was the UFO that the young man was forced to take evasive action. He slammed on the brakes and quickly brought his truck to a screeching halt near the edge of the highway. And there was much worse to come: a second UFO, which was very similar in appearance, landed approximately 50 feet away, and from which two very tall humanoid beings suddenly exited. The man sat and stared in horror as the pair extended their arms, pointed their fingers in the direction of the terrified young man, and began to stride purposefully toward him. Fortunately, albeit somewhat bafflingly, the alien pair decided to break off at the last moment. They returned to their cosmic vehicle, and were soon airborne and gone. But it would not be long before the aliens found their next victim to terrorize.

Barely a couple of hours later, once more in the vicinity of Valdehijaderos, and yet again in the dead of night, something truly otherworldly was afoot. Given that this is a case of jaw-dropping proportions, I have elected to quote directly and in full from the official summary provided to NASA by the Department of Defense that details the story of the male witness:

> Three silver ships parked on the highway with light similar to floodlight. Observer stopped motor of his car and some figures approached him. He ran, frightened, and they followed him. He threw himself into a gutter. His pursuers passed within 2 meters and he saw them. They were about 2 meters tall, had arms and legs but he did not see their faces. After they passed he returned to the truck. The beings returned to observe him again, then they entered their ships and left.[3]

Might this possibly have been some form of attempted alien abduction? Now, 36 years later, it seems impossible to say with any degree of certainty.

A very strange, and perhaps unique UFO event occurred over Malaga one day later, involving the sighting of a large, illuminated "gaseous mass." Whatever its nature and intent, it provided no clues to the startled witnesses.[4]

One of the most credible reports that reached NASA via the Department of Defense surfaced on April 14, 1974, involving three witnesses: a well-respected university professor and his wife (who hailed from the Herrera de Alcantara region of Spain), and one of the professor's students. The trio witnessed, low in the skies, a 65-foot rhomboid UFO that was brightly illuminated by a combination of pink and yellow lights. After a few minutes the craft gently began to descend, finally reaching ground level, where it remained for around five or six minutes before returning to the skies and ultimately becoming lost to view as it headed in a northeasterly direction.

Not even a day had passed by after the encounter of the university professor, his wife, and student at Herrera de Alcantara, when dozens of passengers aboard a ferry en route to Algeciras from Ceuta were the next to be deemed worthy of potentially nonhuman visitation. The passengers were unanimous in their reports: Some form of a very brightly lit object had risen out of the waters near a large rock, had flown by both silently and slowly for a few moments, had descended back into the dark waters, and had repeated the same, precise motions once more, but then never again resurfaced.

CONFIDENTIAL

For the next two months, NASA learned, the UFO hysteria that had gripped Spain out of the blue suddenly subsided. It returned with a vengeance, however, on June 16, 1974. Caceres was the location, the time was around 5 a.m., and the solitary witness was a laborer driving to work. It was a morning he was destined never to forget. As in several of the earlier Spanish cases that attracted the attention of NASA, the UFO chose a stretch of highway to put in its weird appearance: It swooped down on the man's vehicle, chased the terrified soul at a high rate of speed for several miles, and came so close that the driver could see three, giant beings situated inside the craft. Perhaps acting on sheer desperation, the man turned off his vehicle's headlights, at which point the UFO suddenly vanished. Of course, any attempt to drive on a stretch of highway without the benefit of headlights would be a danger-filled experience at the best of times, never mind when being chased by a UFO piloted by Goliath-like extraterrestrials, and so when he felt that the UFO had moved on, the man turned his lights back on to full illumination. This would prove to be a very big mistake: The

UFO suddenly reappeared, seemingly out of thin air, and again came way too close for comfort to the man's vehicle. Of much greater concern, the man reported that the UFO followed him all the way to his home, after which it shot away into the heavens. Fortunately for the witness, it never returned.

And there end the reports. Interestingly, NASA was advised by the Department of Defense in its closing words on the Spanish wave of UFO encounters of 1973–74 that:

> ...in April of this year teams of extra sensory perception specialists held a meeting in Malaga for the purpose of scientifically studying the UFOs seen in that vicinity. [The] results of this meeting [are] unknown.[5]

CLASSIFIED

Moving away from Spain, one of the very strangest—and most disturbing, as a result of its defense implications—of all cases that reached NASA from this same time period originated from within a facility at Frimley, England. It was a facility that fell under the auspices of GEC-Marconi, whose Marconi Underwater Systems, Ltd., and Marconi Space & Defense Systems played leading roles in military-based research and development programs for the British Ministry of Defense.

The date was early 1974, and at the time the primary source of the story was employed as a draughtswoman in the Central Services Branch, having previously served an apprenticeship in Britain's Royal Navy. On arriving at work on one particular morning she was surprised to see numerous British Ministry of Defense personnel swarming around the building. Although the woman was well aware that something of significance had clearly occurred, it was not until later that she was able to approach a trusted colleague who was himself a manager at the installation.

"Something very serious has happened, hasn't it?" she inquired, in hushed tones.

"Yes," her colleague advised, in an equally quiet fashion. "We've had a break-in. I can't say anymore." Throughout the course of several weeks, further pieces of the puzzle began to fall into place. It transpired that the "break-in" was far more than simply an unauthorized entry. What occurred was nothing short of the penetration of a highly guarded facility by a living, breathing extraterrestrial entity.

The incident had occurred late at night, and the primary witness was a security guard who had been patrolling the building as part of his routine duty. While walking along one particular corridor in the installation, the guard was startled by a dazzling blue light emanating from within a nearby room. But this was no ordinary room—it was a storage facility for top-secret documentation generated by Marconi as part of its work on behalf of the British government, much of which was related to classified radar-related defense projects.

Realizing that no one should have been in the room at that time of night, the guard burst in, only to be confronted by a shocking sight. There, sifting through pages of classified files was a humanoid—but decidedly nonhuman—creature that quickly dematerialized before the eyes of the shocked guard. Although severely traumatized by the event, the man was able to give a brief description of the creature to his superiors, and noted that the blue light issued from the front of a large helmet that covered the entire head of the entity.

By the following morning, the guard had suffered a near complete nervous collapse and was taken, under military guard, to an unspecified hospital for intensive therapy.

Some weeks later, the draughtswoman had occasion to overhear snippets of a conversation that took place in the office of her superior, a Mr. Bevan, in which major concern was expressed about the ability of these unidentified creatures to access highly sensitive installations and data. The matter was quietly closed, with a distinct hope on the part of all involved that it would not be repeated.

TOP SECRET

These many and varied profound accounts offer demonstrable evidence of deep NASA interest in UFO reports of a particularly startling nature in the early to mid 1970s. And you should not be surprised to learn that NASA's interest in UFOs, alien life, and related issues of a cosmic and space-borne nature was not destined to fade away in the slightest.

Chapter 8
The Space Brothers

Since the latter part of the 1940s, hundreds of individuals all across the planet have asserted they have experienced face-to-face contact with very human-like aliens from faraway worlds. The particular extraterrestrials at issue are most often described as being attired in tight-fitting one-piece outfits, with heads of luxuriant, flowing hair, usually (but not exclusively) blond in color. Not only that: our cosmic visitors inform those of us whom they have identified as being worthy of face-to-face contact that they are deeply worried by our aggressive nature. It is their desire that we deconstruct our atomic bombs and weapons of mass destruction, that we choose to live tranquil lives, and that our mindset become dominated less by violence and warfare, and instead by thoughts of love and peace. These aliens are now commonly referred to as the Space Brothers, and those people whose existences have been radically altered by their experiences with the Space Brothers are termed the Contactees.

If the words of the eyewitnesses are true, then shortly after UFOs began to appear all across the planet in the summer of 1947, our cosmic visitors chose an alternative approach to their means and method of contact. Supposedly not having any desire to land on the lawn of the White House, outside Buckingham Palace, or in front of the Kremlin, the Space Brothers seemingly preferred a far more direct, and much less sensational means of contact. Deserts, mountains, forests and even, so the eyewitnesses claim, remote restaurants in dusty, windswept little towns in the middle of nowhere were all the preferences of the Space Brothers.

TOP SECRET

One such typical (and nearly definitive) case took place not long after midnight on October 28, 1973. Dionisio Llanca, who was working as a truck driver at the time, was traveling along Highway 3 near to Bahia Blanca, a city situated in the southwest of the province of Buenos Aires, Argentina. For several hours, a slow tire puncture had been affecting Llanca's driving, but a sudden loss of nearly all its remaining air nearly resulted in Llanca losing control of his truck and its cargo. As a result, he was quickly forced to bring his vehicle to a rapid rest at the side of the road, and began to remove the flat tire. Then something very strange happened: Llanca suddenly found himself bathed in a bright beam of light coming from a circular-shaped craft that was hanging in the sky at a height of no more than 20 feet. Partly frozen to the spot by the powerful illumination, Llanca was utterly astounded when, upon turning around, he saw standing to his rear a trio of aliens, all with their eyes burning into his very soul.

The trio, composed of a pair of very human-looking males and one female, were all about 5 1/2 feet tall, and were dressed in snuggly fitting, one-piece, pale, gray outfits. They sported the obligatory blond hair too. The rest of the experience was not a fun one for Llanca. One of the male aliens suddenly lunged violently in his direction, and some unknown contraption was attached to one of the fingers on his left hand. Whether due to fear or as a result of the actions of the device itself, Llanca felt suddenly dizzy, and he soon passed out on the spot. Shortly afterward, and upon finally regaining his senses and his equilibrium, Llanca's mind was flooded with a message from his three visitors from beyond, who told him that they had been present in our world since 1950. They also said, in stern tones:

> [Your] planet is bound to suffer very grave catastrophes if [your] behavior continues as it is at present.[1]

CLASSIFIED

It so transpires that on October 29, 1973, only one day after Dioniso Llanca's amazing experience in Buenos Aires occurred, P.T. McGavin had his own sensational encounter of the Contactee kind. (McGavin was formerly an employee of an outfit with which NASA had contracted on its Gemini program, before taking early retirement at the age of 48 to run a bed and breakfast establishment with his wife just outside of Denver, Colorado.) McGavin's encounter did not take place anywhere near to that of Llanca, however. Rather, it was deep in the heart of New Mexico.

Situated in what is known today as the Four Corners area, where the states of Arizona, Utah, Colorado, and New Mexico come together, the small New Mexican town of Aztec can be found just short of 200 miles northwest of the city of Albuquerque. Without doubt, the strangest thing about the otherwise innocuous town is its infamous crashed-UFO legend. According to information related to the author and journalist Frank Scully in the late 1940s, and subsequently published in his bestselling 1950 book *Behind the Flying Saucers*, as a result of a number of separate incidents that had occurred in 1947 and 1948, the wreckage of four alien spacecraft, and no fewer than 34 alien bodies had been recovered by U.S. authorities and were being carefully studied under cover of the utmost secrecy at a variety of defense establishments in the United States.

Scully said that the bulk of his information had come from two prime sources: Silas Newton, who was described in a 1941 FBI report as being a "wholly unethical businessman," and a "Dr. Gee," a name supposedly created to collectively protect the identities of eight scientists, all of whom, it was reported, divulged various details of the crashes to Newton, and, subsequently, to Scully. According to Scully's alleged sources, a crashed UFO, containing no less than 18 dead, alien dwarfs, was found atop a mesa in Hart Canyon, Aztec, in March 1948. Inevitably, taking into consideration the fact that the bulk of the data on the Aztec affair came from the ethically questionable Newton, the incident was championed by some UFO investigators, such as Bill Steinman and Wendelle Stevens in their 1987 book *UFO Crash at Aztec*, and utterly denounced and hammered into the ground by others, including the late Karl Pflock.

To this day, the story of Aztec's crashed UFO continues to court controversy. The UFO research community remains very much divided on what did or did not occur on that long-gone day. To some, the affair is evidence that aliens really did crash at Aztec. For others, it is nothing but a hoax that has now practically taken on a life of its own. However, the events of March 1948—whatever their ultimate nature—were not the only ones that suggested alien entities had paid a visit to Aztec.

Like many of the early and formative Contactees, prior to his startling encounter of 1973, P.T. McGavin (whom I met in 2003 while I was lecturing at Aztec's annual UFO conference, which he attended) had no prior interest in or fascination with UFOs, aliens, or extraterrestrial contact. Yes, his contract work with NASA's Project Gemini did, of course, ensure that much of his career was dominated by the realm of outer space. In his own words, however:

> We were just looking to get the Gemini boys doing the best they could. Aliens really weren't on my mind back then.[2]

It's safe to say, however, that after October 29, 1973, UFOs and alien life were seldom absent from McGavin's mind.

It was around 9 a.m. on October 28, said McGavin, when he began to feel extremely restless, but for reasons he could not readily explain. The whole day prior to that, October 27, had been a somewhat hectic one at the bed-and-breakfast establishment, and he at first wondered if he was merely experiencing a case of "fried nerves."[3] As the day progressed, however, the restlessness worsened, and then something very strange happened: a disembodied female voice, speaking in near whispered tones, said:

> Take yourself to Aztec, New Mexico. Your future is ours, too. We welcome you.[4]

Petrified out of his wits that he was perhaps exhibiting the first signs of mental illness or even a brain tumor, McGavin paced around the room, utterly unable to relax. Three more times during the course of the next 20 minutes or so, the voice repeated the exact same words, which, not surprisingly, quickly became ingrained in McGavin's memory. Then, without warning, stark visual imagery of a vast, black-colored, triangular-shaped UFO hovering over what he would later come to realize was Aztec's Hart Canyon filled his amazed and frenzied mind. Whatever was taking place, it was not remotely like anything McGavin had ever previously experienced. It was then that McGavin chose to do something that "sounds crazy looking back, but I knew I had to do it, and I knew it had to do with aliens."[5] That something was a road trip to Aztec, while his wife valiantly coped with keeping the bed and breakfast running.

The oddest thing of all about the initial stages of the experience, reflected McGavin, was that his wife seemed to take the whole thing in stride and exhibited no concerns whatsoever about the fact that her husband was about to take an extended drive to a small New Mexican town, solely on the words of a disembodied voice. Indeed, McGavin later wondered, but was very careful not to tell his wife, if she had been somehow temporarily "programmed" by the aliens "not to worry," and to ensure that the operation ran smoothly and on schedule.[6] As a result, McGavin filled his car with gas, and, around 2 p.m., hit the road. Aztec and the aliens were both demanding his attention.

CONFIDENTIAL

Aside from brief stops to refill his car with gas and to grab a sandwich, snacks, and drinks for the road, he drove the whole 350-mile-plus journey in one stretch, arriving in Aztec around 9 p.m. Notably, although much of the town had shut down by then, he inquired at a nearby diner if anyone knew anything about UFO activity in the region. Well, of course they did: the legend of the 1948 UFO crash at Hart Canyon, regardless of whether they actually believed in the story or not, was known to practically everyone in town. Having secured directions to the alleged crash site from a waitress in the diner, he followed both the map and his instincts and made his cautious way to the mesa where aliens from another world allegedly met their deaths a quarter of a century earlier. The last part of the journey, said McGavin, was undertaken on foot, due to the fact that driving his car up to the mesa was simply not a viable option—at least, not if he wanted his car to remain intact and drivable for the return journey. So, with a powerful flashlight in hand, he made his careful way up to the top of the mesa, sat down, and waited. And waited.

By the time midnight arrived, McGavin's patience was wearing decidedly thin. But, he felt the need to stick it out—after all, he had traveled more than 350 miles to come here, and so leaving now would be a major mistake, he considered. It's fortunate that he did decide to bide his time and see what happened. For a short period, said McGavin, he fell asleep while stretched out on the backseats of his car, but around 2 a.m., he was awakened by a deep, intense humming sound that resonated throughout his whole body and which made him feel distinctly nauseous. The stomach-churning sickness was the least of McGavin's worries, however. As he wound down one of the back windows of his car and looked skyward, he was shocked to the core to see barely 80 feet above him a large, triangular-shaped, black-colored UFO that was practically identical to that which he had seen at the height of his visionary experience early on the previous day.

He flung open the car door, and as he did so, the nausea and the pounding humming noise both suddenly vanished. All around him there was silence. Three small lights issued forth from below the circular craft and offered at least some illumination. If seeing a UFO was not enough for McGavin, what happened next was even more unbelievable. Only about 30 feet to his right, he saw a humanoid figure slowly walk out of the darkness toward him. Unable to move or speak, McGavin merely stared as the figure got ever closer and was then lit up by the lights of what McGavin was now convinced was nothing less than a spacecraft from another world.

The figure very much resembled a human male, close to 6 1/2 feet in height, dressed in a one-piece, gray-colored outfit that extended from halfway up his neck down to his feet, and sported a head of long, shining blond hair that nearly reached down to his waist. When the being held out his hand, McGavin's first, and quite understandable instinct was to make a quick run for it. Instead, he took a deep breath and extended his hand. A firm handshake followed, after which the alien offered a slight smile, and introduced himself as Gavon.

TOP SECRET

According to McGavin, despite his deep hopes, he was never taken onboard the UFO, nor was he ever informed of the precise point of origin of his mysterious visitor. Rather, as the UFO continued to hover overhead and illuminate the surroundings, Gavon motioned McGavin to an area of dusty ground, promptly sat down cross-legged, and invited McGavin to do the same, which he did. As McGavin listened carefully, Gavon told him—in a voice that seemed to have a slight, but distinct female quality to it—that he and others of his kind had been visiting the Earth on a clandestine basis since the outbreak of the Second World War. The aliens had even chosen to live among us, which, taking into consideration Gavon's very human appearance, would not have made such a thing impossible, said McGavin, even though one might look at Gavon twice, primarily as a result of his height and long blond hair. At the time of the hostilities—1939 to 1945—the aliens had restricted their actions to those of observers only. With the development of the atomic bomb, however, and with the subsequent destruction of the Japanese cities, Hiroshima and Nagasaki that quickly ended the Second World War, Gavon said his people took a decision to drastically change their tactics. They decided to become far more proactive, and just about everywhere.

Beginning in the summer of 1947, the being told McGavin, the aliens began to discreetly contact human beings all across the planet. There was no rhyme or reason to the contact; it was all entirely at random. There was, however, a method behind this seemingly illogical approach: Contacting people solely on a chance basis lessened the possibility of the governments and military agencies of several of the leading nations (which were then starting to take notice of the UFO mystery) from tracking them down and capturing them. In reality, said Gavon, the alien presence was actually a very small one. There had never been more than 40 of his people and eight of their craft, either on or orbiting our planet in the 1940s and 1950s, but the use of sophisticated holographic-style imagery and stage-managed encounters had successfully instilled in the minds of many the scenario of a powerful, veritable armada of alien craft visiting our world on practically a daily basis. That the real alien presence was actually quite miniscule, said Gavon, required a degree of significant deception on their part to ensure that we, as a race, concluded the aliens were all-powerful. In reality their numbers were small, their craft could hardly be described as a fleet, and they were a long way from home, operating with the benefit of extremely limited resources.

Of course, the biggest question facing McGavin was: Why had he been chosen? Gavon simply smiled again at first, and then explained that he and his kind wished to subtly get a message across to the people of Earth—one of love, of light, of end to the Cold War, and of the creation of a new world that would see the human race elevated to a whole new level. And, in much the same way that they had done with some of the Contactees of the 1950s who claimed very similar experiences to McGavin (such as George Van Tassel, Orfeo Angelucci, and Truman Bethurum), Gavon wanted McGavin to speak publicly about his experience atop the mesa, to have people think about and muse upon the encounter, and help further instill in our minds the notion that kindly aliens were among us, and were here to help in our development, scientifically and spiritually, as a species.

Seemingly very interested in gauging McGavin's opinion on such matters, Gavon stared intently in his direction, in complete silence, for about 20 seconds. This earsplitting silence, coupled with the eerie stare from a being that claimed to be an emissary from another world, greatly disturbed McGavin, who, to break the silence, said that he fully understood the profound importance of what was being requested of him, and would follow the path that Gavon had chosen for him. The long-haired alien stood up, seemed very satisfied with the outcome of the brief exchange that the pair had, and retreated into the surrounding darkness until he was lost from view. Curiously, said McGavin, he never saw Gavon enter the UFO; however, after a few minutes it rose slowly, and in complete silence, into the sky, at which point its lights were extinguished and it was lost to McGavin's sight. The experience was over, and unlike many other Contactees, McGavin was never again blessed with a visit from the stars. All was not well with McGavin, however.

CLASSIFIED

On the drive back to Denver, Colorado, he could not shake the disturbing feeling that he had just been brilliantly used and exploited by a being that was wholly deceptive in nature and did not have our best interests at heart. There was something about Gavon, McGavin recalled, that was not only too good to be true, but was also very unsettling, although he admitted he could not quite place what it was that he found so disturbing. There was also the fact that the name of the alien, Gavon, was not that dissimilar to McGavin's own last name. Was this merely due to a bizarre coincidence or synchronicity, or was Gavon trying to create a bond between the two by using a spurious identity? McGavin had no way of knowing, but he was not at all happy with the situation.

Throughout the course of the next few weeks, McGavin pondered deeply upon the experience and finally chose to take what, with the benefit of hindsight, may have been a big risk: he contacted several of his old friends from NASA's Gemini program and spilled the beans on what had occurred that night in Hart Canyon. As a result, perhaps six weeks afterward, he received a telephone call from a man identifying himself as a Mr. Callanan, who said he worked with NASA on "things to do with security, and asked if we could meet somewhere I could choose."[7]

Evidently, travel was not a problem to Mr. Callanan, and the pair met three days later in a restaurant near Denver's main airport. Callanan explained to McGavin that the details of his experience had been brought to his attention via one of those former Gemini operatives that McGavin had spoken with about the encounter—"They all denied it later," McGavin told me.[8]

In serious yet also sensational tones Callanan told the worried McGavin that he was absolutely right to have had deep seated concerns about his strange conversation with Gavon. Despite what McGavin had been assured by his supposedly cosmic visitor, Callanan advised him that Gavon and his small band of comrades were not aliens from some distant galaxy at all. Rather, they represented the last vestiges of a very ancient and very terrestrial race of beings that were closely related to the human race, that—tens of thousands of years ago, had an advanced but isolated civilization that was responsible for the legends of Atlantis and similar stories, but who were forced to retreat into huge, carved, underground caverns when they were faced with an overwhelming, ever-growing, violent problem that was quickly infesting the whole planet. That problem was us: the human race, said Callanan.

The concern that Gavon expressed that we, as a species, might very well irreparably damage the Earth via a disastrous nuclear exchange between NATO and the Warsaw Pact was very real, McGavin was told. This was no lie, explained Callanan. But that fear was borne out of the fact that Gavon, and all of his kind, were forced to share the planet with us—something that, for McGavin, made a great deal of sense. Of course, numerous questions danced around McGavin's mind: how did NASA know this? Were these ancient humanoids dangerous? What was the full nature of their agenda?

Callanan replied that NASA had secured hard evidence of the existence of these beings, and had been able to pretty much confirm their point of origin—right here, rather than from way out there—in 1968, when two of Gavon's people had been captured following the crash of one of their craft, which proved to be a strange, dirigible-type craft, in the wilds of the New Mexico desert. Silent for months, the long-haired pair, after being subjected to what Callanan enigmatically described as "certain tactics that needed to be done to get the job done,"[9] the dark truth finally came tumbling out: our presumed extraterrestrials were nothing of the sort. Rather, they had been here all along, trying to find a way to take back the world that had once been theirs, and extinguish their most hated foe—us—once and for all in the process.

The problem was that their species really was, by now, Callanan explained, very impoverished, and on a definite wane from an evolutionary perspective: genetic problems and general ill health had forced this humanlike race to try and interbreed with us, as a means to try and improve their stock and save them from the icy clutches of extinction. And, as there appeared to be an ancient lineage that suggested we shared a common point of origin, the crossbreeding was actually working, much to the chagrin of NASA. But, to ensure they remained free of detection, the beings constructed false stories that they were from distant star systems.

The U.S. government, Callanan said, was mightily concerned about the situation, and the only thing of any real significance in our favor was that although the beings still possessed elements of their original, highly advanced technologies, they were certainly not on a scale that would allow them to go to war with us, en masse. Their process of elimination of us, it was concluded by certain elements of NASA, was to bide their time and figure out a way to remove us from the face of the planet in another way—perhaps by the use of a lethal virus that would prove to be deadly to us, but to which they would have complete immunity (how NASA had reached this conclusion was not made clear to McGavin). McGavin had done the right thing by relating his experience to his friends who had worked on NASA's Gemini program, said Callanan, and everyone on the UFO project was grateful—although certainly not happy—to know that these ancient creatures were still spreading their alien deception.

After the cloak and dagger meeting with Callanan, McGavin's head was swimming. Was the story true? Or was it a cover story designed to confuse the possibility that kindly aliens, which the government did not wish to see radically change the nature of our society, really were among us? When I met with him in 2003, McGavin admitted that he still had no answers to those questions, primarily because any and all attempts to locate the mysterious Callanan utterly failed, and he never again crossed paths with anyone else in NASA who knew of or was willing to discuss such matters.

CONFIDENTIAL

Notably, there are those who share (or shared) the views and conclusions of Callanan. One of those was the late Mac Tonnies, whose posthumously published book, *The Cryptoterrestrials*, focused upon the idea of our alien visitors really being the last, dwindling remnants of an ancient Earth-based culture.

In 2009, Tonnies told me that, with respect to the long-haired, blond aliens and those they chose to contact:

> Commentators regularly assume that all the Contactees were lying or else delusional. But if we're experiencing a staged reality, some of the beings encountered by the Contactees might have been real; and the common messages of universal brotherhood could have been a sincere attempt to curb our destructive tendencies. The extraterrestrial guise would have served as a prudent disguise, neatly misdirecting our attention and leading us to ask the wrong questions—which we're still asking with no substantial results.[10]

Tonnies added:

> Contactees and abductees alike describe the interiors of 'alien' vehicles in curiously cinematic terms. The insides of presumed spaceships often seem like lavish props from never-filmed sci-fi dramas. The aliens don't fare any better; they behave like jesters, dutifully regurgitating fears of ecological blight and nuclear war but casually inserting

allusions that seem more in keeping with disinformation than genuine E.T. revelations.[11]

In other words, Mac Tonnies, just like Mr. Callanan of NASA, came to accept that the long-haired aliens of Contactee lore were not just deceptive creatures; they were definitively home-grown aliens.

TOP SECRET

There is another matter of deep relevance to this whole controversy that is worthy of comment too, and which has a major bearing upon McGavin's curious encounter, as well as upon the nature of the UFO he saw at Hart Canyon, Aztec: the huge, black, triangular-shaped craft.

For many people, any mention of the word *UFO* inevitably conjures up imagery of flying saucers of the type that have dominated popular culture since 1947, during the summer of which the term was coined. The reality of the situation, however, is that UFOs come in all manner of shapes and sizes. And, for the last 20 years or so, sightings of craft of unknown origins very similar to that seen by McGavin in October 1973 have proliferated within the annals of UFO research. Unsurprisingly, they have become known as Flying Triangles, or FTs. For some UFO researchers, the Flying Triangles are perceived as being highly classified military vehicles, something very much along the lines of a next-generation Stealth-style aircraft. Other researchers, meanwhile, are not quite so sure. If such reports were purely the domain of the late 1980s onwards, then a very strong and logical case could be made that this new addition to the UFO puzzle had indeed just rolled out of a secret aircraft hangar somewhere in the deserts of Nevada. Accounts such as that of McGavin, however, demonstrate that the Flying Triangle mystery is far older than many investigators of the phenomenon have previously assumed possible.

Furthermore, McGavin's story is far from being alone in terms of early Flying Triangle reports. For example, a now-declassified British Ministry of Defense file of March 28, 1965, briefly describes the experience of a witness who saw a veritable fleet of such triangular craft flying over remote moorland in North Yorkshire, England, on the previous night. According to the document, the witness, one Jeffrey Brown, saw "Nine or 10 objects—in close triangular formation each about 100 feet long—orange illumination below—each triangular in shape with rounded corners, making low humming noise."[12]

The low humming noise, the triangular shape, and the illumination coming from below the craft are nearly identical characteristics to integral aspects of the craft that McGavin viewed on the other side of the world almost a decade later. And while we are on the subject of the British Ministry of Defense, the following case is a perfect example of a latter-day encounter with a similar Flying Triangle that attracted official interest on the part of British authorities.

CLASSIFIED

The events began on the night of March 31, 1993, when numerous people all across the British Isles were witness to unusual lights and craft in the skies of the country, and who chose to contact either local police forces or the Ministry of Defense to report the details of their encounters with the unknown. Of the many and varied reports that reached the Ministry of Defense, what was surely the most dramatic and credible came direct from none other than a serving military source. Nick Pope, now retired from the Ministry of Defense, was tasked with investigating this particular case at an official level at the time of its occurrence. I had the good fortune to interview Pope on this matter in March 1998, and he imparted extraordinary data to me, on the case, the witness, and his, Pope's, personal conclusions regarding what the affair represented.

The location was a Royal Air Force base called Shawbury, which is situated in the central England county of Shropshire; the witness was a meteorological officer with the military; and the overall event was destined to have a remarkable effect on Nick Pope himself.

Pope said to me, when I brought up the matter of the Shawbury encounter:

> Military officers are very good at gauging sizes of aircraft, and they're very precise. [The witness's] quote to me was that the UFO's size was midway between that of a C-130 Hercules and a Boeing 747 Jumbo Jet. Now, he had eight years' worth of experience with the Royal Air Force, and a Meteorological Officer is generally much better

> qualified than most for looking at things in the night sky. And there were other factors too: He heard this most unpleasant low frequency hum, and he saw the craft fire a beam of light down to the ground. He felt that it was something like a laser beam or a searchlight. The light was tracking very rapidly back and forth and sweeping one of the fields adjacent to the base. He also said—and he admitted that this was speculation—that it was as if the UFO was looking for something. Now, the speed of the UFO was extremely slow, no more than 20 or 30 miles per hour, which in itself is quite extraordinary. As far as the description is concerned, he said that it was fairly featureless; a sort of flat, triangular-shaped craft.[13]

Perhaps the most revealing aspect of the RAF Shawbury encounter of March 31, 1993, was the way in which the Flying Triangle made its exit, as Nick Pope explained to me:

> [The Meteorological Officer] said that the beam of light retracted into the craft, which then seemed to gain a little bit of height. But then, in an absolute instant, the UFO moved from a speed of about 20 or 30 miles per hour to a speed of several hundreds of miles per hour—if not thousands. It just suddenly moved off to the horizon and then out of sight in no more than a second or so—and there was no sonic boom.[14]

Despite the fact that the Ministry of Defense launched an extensive investigation, the matter of the RAF Shawbury Flying Triangle was never resolved. It was an important incident, for Nick Pope:

> I don't know if it was the single turning point that switched me from being an open-minded skeptic to a

believer, but it was certainly one of the key events. In fact, if you were to ask me to take my best shot, I would say that this was the real article; this was extraterrestrial.[15]

CONFIDENTIAL

Was Pope's conclusion that the Flying Triangles are of alien origin correct? Or has NASA chosen to keep British authorities in the dark about its beliefs and conclusions relative to the Flying Triangles being connected to the deceitful activities of an ancient, terrestrial race? We don't have the answer to those questions in hand just yet, but it is intriguing to note that in 1999 Nick Pope wrote a science-fiction novel titled *Operation Thunder Child*, which told of an alien invasion of the Earth from the perspective of the Ministry of Defense and the British government. Notably, in Pope's novel, the U.S. government tries to deceive the British into believing that the aliens are an offshoot of Neanderthal Man that quite literally went underground millennia ago, "developing their own complex social structures and technologies."[16]

At one point in Pope's book, the U.S. president confides in the British prime minister that these ancient humans "have been keeping a careful watch on our development, especially since the Industrial Revolution. But it's our progress in the last hundred years that has most frightened them." Of course, this is all very reminiscent of what the purported alien Gavon told P.T. McGavin in October 1973. The only major difference is that NASA believed that the extraterrestrial angle was a deception to hide the fact that the purported aliens were actually ancient beings from our own Earth, and in Nick Pope's novel, the situation is exactly the opposite: the ancient human angle is a cover story to mask a real alien presence.

TOP SECRET

This all, inevitably, provokes a couple of notable questions: did Nick Pope, as a former UFO investigator for the Ministry of Defense, hear whispers and rumors—from American friends and colleagues in the world of officialdom—of the theory that our alien visitors may not be from faraway star systems after all, but instead might be home-grown? And, if so, did he then choose to weave certain aspects of this scenario into his novel? We cannot say for certain. It is important to note, however, that one year before *Operation Thunder Child* was published I questioned Nick Pope vigorously regarding the growing claims and rumors that his novel was going to be based far more upon secret fact than entertaining fiction. His comments were highly illuminating, to say the very least:

Pope told me:

> Even to you, Nick, I can't comment on that. But let's put it this way: Operation Thunder Child is going to be more controversial than Open Skies, Closed Minds or The Uninvited [Pope's two previous nonfiction titles on UFOs that were published in 1996 and 1997, respectively]. And, indeed, the Ministry of Defense may have more of a problem with it. Mainly because it's going to feature real locations, real weapon systems, real tactics, real doctrine, and real crisis-management techniques. It's going to blend my knowledge and experience of UFOs with my knowledge of crisis management—such as my involvement in the Gulf War where I worked in the Joint Operations Center.[17]

I continued with my questions: "Given that you won't comment on the claims that are circulating right now that *Operation Thunder Child* will relate in a fictional format the sorts of things that you were legally unable to relate in a non-fiction book, are you saying that there is more going on behind the scenes?"

Positively oozing uneasiness, he replied:

> Well, it's very difficult to go into the details, but I'm a bit more inclined to think that there's perhaps more to this than meets the eye.[18]

Based on all that we have digested in this particular chapter, I cannot disagree with Nick Pope in the slightest.

Chapter 9

Mac Tonnies Faces the Martians

Located in an area of the planet Mars known today as Cydonia, bearing an uncanny similarity to a human visage, is a huge structure that lies 10 degrees north of the Martian equator. It was first photographed on July 25, 1976, by NASA's *Viking 1* space probe, which was orbiting the planet at the time, and was brought to the attention of the public in a NASA press release on the photograph six days later. It has since become universally, and infamously, known as the Face on Mars.

It is a fact that most interpretations of the photograph offered by both NASA and by certain elements of the mainstream scientific community suggest that the face is nothing more than an entirely natural landform, one of the numerous mesas that certainly can be found throughout

A panoramic view of the location of the famous Face on Mars.

the entire Cydonia region. But far more intriguing is the theory that the face might very well represent the ruined remains of an ancient, artificial sphinx-style monument of some kind, possibly one that was built by an indigenous but now very long extinct Martian civilization. In 1987, Richard Hoagland, who has dug deeply into the controversy surrounding the Face on Mars, noted that large formations, not unlike the pyramids of Egypt, could be seen in close proximity to the face. Hoagland, whose book *The Monuments on Mars* detailed his findings, suggested that such formations might offer further evidence that Mars was once inhabited by intelligent beings.

Photographic evidence secured from NASA's *Mars Global Surveyor* probes in both 1998 and 2001, and then by the *Mars Odyssey* probe in 2002, disappointed many who had previously championed the idea or theory that the face was an ancient, artificial construction. The new pictures looked like just about anything but a face. Perhaps inevitably, and possibly with some justification, conspiracy theorists promptly claimed that the always mysterious "they" had carefully altered the photographs, in a deliberate, top-secret effort to make them all appear far more mesa-like and, of course, far less face-like. Years later, the controversy continues to rumble steadily. To fully understand and appreciate the nature of this curious and potentially Earth-shattering (or maybe *Mars-shattering* would be a better and more accurate description) affair, we have to focus our attention upon a particularly notable individual who carefully studied, and commented on, the Face on Mars controversy for years.

TOP SECRET

Perhaps more than anyone else, one person had a deep understanding and appreciation of the vital need to remain both unbiased and balanced on what the Face on Mars does or does not portray or represent: the late, and previously mentioned, Mac Tonnies, an industrious, enthusiastic, and careful researcher, lecturer, and author, who passed away in October 2009 at the age of only 34. His 2004 book, *After the Martian Apocalypse*, arguably remains the most carefully reasoned and definitive piece of work on this highly controversial and divisive matter. I had the fortunate opportunity to interview Tonnies on a number of occasions—in 2004, 2005, 2006, and 2007—and I believe his words and conclusions pertaining to the Face on Mars still stand up to scrutiny to this very day.

He said to me:

> I've always had an innate interest in the prospect of extraterrestrial life. When I realized that there was an actual scientific inquiry regarding the Face and associated formations, I realized that this was a potential chance to lift SETI—the Search for Extraterrestrial Intelligence—from the theoretical arena. It's within our ability to visit Mars in person. This was incredibly exciting, and it inspired in me an interest in Mars itself—its geological history, climate, et cetera.[1]

But when, how, and under what particular circumstances did the controversy really start? Tonnies replied, to the point: "When NASA dismissed the face as a 'trick of light.'"[2]

He elaborated to me:

> NASA itself actually discovered the Face; of course, it was written off as a curiosity. Scientific analysis would have to await independent researchers. The first two objects to attract attention were the face and what has become known as the "D&M Pyramid," both unearthed by digital imaging specialists Vincent DiPietro and Gregory Molenaar.[3]

Of a second image of the face that was located by the two, who had been under contract to NASA's Goddard Space Flight Center, Tonnies said:

> Viking did manage to capture another image of the face, but only days later. Far from disputing the face-like appearance, it strengthened the argument that the face remained face-like from multiple viewing angles. The prevailing alternative to NASA's geological explanation—that the face and other formations are natural landforms—is that

we're seeing extremely ancient, artificial structures built by an unknown civilization. It was never publicized, perhaps for understandable reasons.[4]

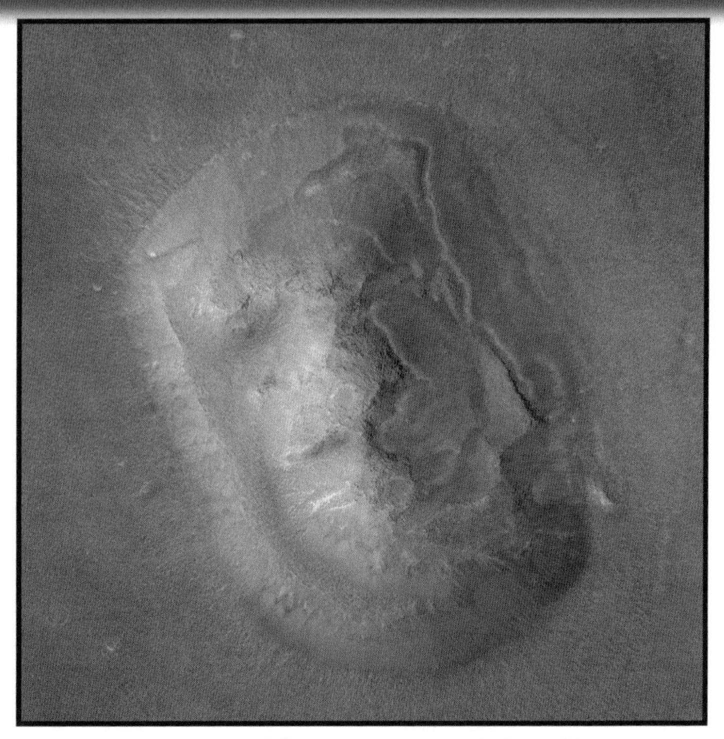

A close-up image of the Face on Mars, believed by many to have been carved by an ancient Martian civilization.

Taken at a more revealing sun-angle than its predecessor, Viking frame 70A13 not only confirmed the salient facial resemblance noted by Owen and Soffen, but showed a continuation of the "mouth" feature, and, despite apparent odds, a second "eye." In any case, NASA had never correctly interpreted the image as an optical illusion caused by the illumination angle of the sun, having never investigated. And if the facial likeness was merely an illusion, why does it persist in more recent images? One would rightly expect a natural surface

> formation to look less like a face when seen in high resolution.
>
> While the Face, natural or otherwise, exhibits the erosion and degradation expected of a mile-long morphology—especially on the eastern side, which appears to have collapsed inward under a hefty layer of accumulated debris—it features secondary facial detail consistent with the impression of an artificial construction. Most notably, the western "brow" shelters an anatomically correct "iris," properly positioned within a unique almond-shaped basin. And at least one "nostril"—never actually visible in the original Viking image—is plain to be seen; while its origin remains an unanswered question, flatly refuting its existence smacks of a deliberate attempt to make the face go away in the public mind.[5]

Tonnies expanded further:

> The fact that virtually no one seriously considered Mars to be home to an extant civilization was brushed aside to accommodate the skeptical community's need to shoot down the looming myth that the face has become in the decades since it was first photographed. Sadly, the opportunity to address the issue of extraterrestrial archaeology in scientific terms was squandered, leaving a residue of misconceptions that only fueled the "fringe's" obsession with conspiracy theories. Fortunately, there's no reason we can't take up the case for unbiased, disciplined appraisal of candidate Martian artifacts.[6]

And what, exactly, did NASA have to say about the controversy of the Face on Mars and the various conclusions of the independent research community that was now beginning to see the structure as one worthy of deep investigation?

Tonnies replied:

> NASA chooses to ignore that there even is a controversy; or at least a controversy in the scientific sense. Since making the face public in the 1970s, NASA has made vague allusions to humans' ability to "see faces" and has made lofty dismissals. But it has yet to launch any sort of methodical study of the objects under investigation. Collectively, NASA frowns on the whole endeavor. Mainstream SETI theorists are equally hostile. Basically, the face—if artificial—doesn't fall into academically palatable models of how extraterrestrial intelligence will reveal itself, if it is in fact "out there." Searching for radio signals is well and good, but scanning the surface of a neighboring planet for signs of prior occupation is met with a very carefully cultivated institutionalized scorn. And, of course, it doesn't help that some of the proponents of the face have indulged in more than a little baseless "investigation."[7]

Tonnies had much to say about the eerie appearance of the face as well:

> Mainstream skeptics commonly dismiss the Face on Mars as an example of "pareidolia," the brain's attempt to attribute meaning to random stimuli. Most of the likenesses described by Face-on-Mars debunkers are profile images. Viewed from only a slightly different angle, the celebrated face-like resemblance vanishes, replaced by an obviously natural phenomenon. While profiles rely on a minimum of information to convey a sense of the mysterious—contours to suggest features

such as a "nose," "mouth," etc.—the Face on Mars is different in several notable respects. For instance, the face appears to be a frontal portrait. While computer modeling reveals a striking facial profile when seen from the perspective of an observer on the Martian surface, the face retains a humanoid likeness when viewed from above. This doesn't prove that the face is the work of intelligence, but it tends to elevate it from the oft-mentioned examples wielded by geologists convinced the Face on Mars must invariably yield to prosaic explanations.[8]

And, notably, Tonnies revealed that:

High-resolution images of the Face reveal detail not visible in the early Viking photographs. Astronomer Tom Van Flandern, for instance, quickly noted the presence of accurately situated features such as an apparent "pupil" in one of the "eyes," as well as "nostrils" and "lips"—all of which were beyond the resolving power of the Viking mission. The low odds of such secondary facial characteristics occurring by chance helped belie the notion that the Face on Mars was the product of garden-variety pareidolia. If the Face on Mars is indeed a windblown butte, it's a great deal stranger than imagined prior to high-resolution scans. Indeed, if the same level of detail had been detected on a terrestrial surface feature, it's probable that archaeologists would have been consulted in order to assess its merit as a potential artifact.[9]

Tonnies continued:

> It would seem the face's unlikely presence on a "dead" world has effectively doomed it to pop science oblivion. But the face is far from a solitary anomaly; it shares the Cydonia region with other, equally intriguing features that call for careful analysis. Taken together, an objective viewer is presented with a gnawing puzzle that may ultimately demolish the easy certainties that coincide with the traditional view of our solar system.[10]

If the formations on Mars—and particularly so the huge face—are indeed artificial, or semi-artificial, then the immensely important question is surely: who built them? Indigenous Martians, or perhaps some visiting ancient civilization that originated from far outside of our solar system in the distant past? Tonnies had his own pet theories and ideas on this particular aspect of the larger controversy:

> It's possible that the complex in Cydonia, as well as potential edifices elsewhere on Mars, were constructed by indigenous Martians. Mars was once extremely Earth-like. We know it had liquid water. It's perfectly conceivable that a civilization arose on Mars and managed to build structures that are within our ability to investigate.
>
> Or, the anomalies might be evidence of interstellar visitation: perhaps the remains of a colony of some sort. But why a humanoid face? That's the disquieting aspect of the whole inquiry. It suggests that the human race has something to do with Mars, that our history is woefully incomplete, that our understanding of biology and evolution might be in store for a violent upheaval. In retrospect, I regret not spending more time in the book addressing the possibility that the face was built by a vanished terrestrial civilization that had achieved spaceflight. That was a tough notion to swallow, even as speculation, as it raises as many questions as it answers.[11]

Did Tonnies personally believe that all of the perceived anomalous and artificial structures are indeed precisely that—anomalous and artificial—or did he conclude that some are of natural origin? His view was both enlightening and logical:

> I suspect that we're seeing a fusion of natural geology and mega-scale engineering. For example, the face is likely a modified natural mesa, not entirely unlike some rock sculptures on Earth but on a vastly larger and more technically challenging scale.[12]

Tonnies had much more to impart to me:

> When the face was reimaged [by NASA] in 1998, debunkers condescendingly noted the lack of "roads" and parked "flying saucers" that would conclusively demonstrate artificiality. But given Mars' age and geological history, superficial features like "roads" would be the last things one might reasonably expect to find; unless, of course, Mars was home to an active alien civilization with a penchant for terrestrial architecture.[13]
>
> Long before the Mars Global Surveyor spacecraft returned provocative images of the Cydonia Mensae region of Mars, the presence of secondary facial characteristics had been predicted by proponents of what became known as the Artificiality Hypothesis. It seemed likely that the Face, if the work of intelligence, would betray traces of anthropomorphic detail when imaged by better cameras. The "eye" was only barely visible in the best of the early Viking photos from the 1970s; certainly little or nothing about its shape or structure could be inferred.[14]

> So when the first overhead images of the face became available, the presence of a seemingly well-preserved "eye" became apparent vindication for proponents of artificiality on Mars. After all, it had been predicted by a testable hypothesis. Other "secondary" features were noted as well: lip-like structures that defined a broad parted "mouth," candidate "nostrils," and others. While none of these features proved that the face was the work of extraterrestrial intelligence—let alone the subject of a far-reaching NASA cover-up—they pointed to the possibility that the face, and perhaps other anomalies in its vicinity, were more than tricks of light, as maintained by NASA's public relations personnel. To date, NASA has yet to conduct a scientific investigation that would bear out its contention that the features in Cydonia are wholly natural—an undertaking that might reasonably include the expertise of archaeologists familiar with the role of remote sensing in detecting potential sites here on Earth.[15]

Adding to the idea that the Martian surface is absolutely brimming with intelligently created curiosities, Tonnies reported that:

> The Mars Global Surveyor has taken images of anomalous, branching objects that look like organic phenomena. [The science-fiction author] Arthur C. Clarke, for one, was sold on the prospect of large forms of life on Mars, and had been highly critical of JPL's [the Jet Propulsion Laboratory's] silence. Clarke's most impressive candidates are what he termed "Banyan Trees" seen near Mars's south pole. And he collaborated with Mars researcher Greg Orme in a study of similar features that NASA has termed "black spiders"; root-like formations that suggest tenacious, macroscopic life. And:

> Mars has water. It's been found underground, frozen. If we melted all of it we'd have an ankle-deep ocean enveloping the entire planet. I predict we will find more of it.[16]

Getting into even deeper and far more tangled controversies, Tonnies reported that:

> There is a superficial similarity between some of the alleged pyramids in the vicinity of the face and the better-known ones here on Earth, such as the Egyptian pyramids and the sphinx. This has become the stuff of endless arcane theorizing, and I agree with esoteric researchers that some sort of link between intelligence on Mars and Earth deserves to be taken seriously. But the formations on Mars are much, much larger than terrestrial architecture. This suggests a significantly different purpose, assuming they're intelligently designed. Richard Hoagland, to my knowledge, was the first to propose that the features in Cydonia might be "arcologies"—architectural ecologies—built to house a civilization that might have retreated underground for environmental reasons.[17]

If Mars was once home to a form of intelligent life, perhaps one not unlike our own right here on Earth, and NASA really did stumble upon evidence of that same advanced intelligence, in the form of the enigmatic, huge face, then how was an apparently thriving planet catastrophically transformed into the utterly dead, dusty, and pummeled world that we see today? And, more importantly, how did its civilization meet its tragic end? Tonnies had some intriguing ideas with respect to these particular questions:

> Astronomer Tom Van Flandern has proposed that Mars was once the moon of a 10th planet that

literally exploded in the distant past. If so, then the explosion would have had severe effects on Mars, probably rendering it uninhabitable. That's one rather apocalyptic scenario. Another is that Mars's atmosphere was destroyed by the impact that produced the immense Hellas Basin. Both ideas are fairly heretical by current standards. Mainstream planetary science is much more comfortable with Mars dying a slow, prolonged death. Pyrotechnic collisions simply aren't intellectually fashionable, despite evidence that such things are much more commonplace than we'd prefer.[18]

On the issue of the Face-on-Mars research community being open-minded or biased as to what the Face may be, Tonnies said:

Frustratingly, this has become very much an "us vs. them" issue, and I blame both sides. The debunkers have ignored solid research that would undermine their assessment, and believers are typically quite pompous that NASA et al. are simply wrong, or, worse, actively covering up.[19]

Did Tonnies personally feel that there is a real, active conspiracy within NASA that requires them to deny the reality of the Face on Mars and the idea that the associated structures are artificial, no matter what?

When NASA and the JPL released the first Mars Global Surveyor image of the Face in 1998, they chose to subject the image to a high-pass filter that made the Face look hopelessly vague," Tonnies explained. "This was almost certainly done as a deliberate attempt to nullify public interest in a feature that the space agency is determined to ignore. So yes, there is a cover-up, of sorts. But it's in plain view for anyone who cares to look into the

> matter objectively. I could speculate endlessly on the forms a more nefarious cover-up might take, but the fact remains that the Surveyor continues to return high-resolution images.[20]

Tonnies was careful to stress that:

> Speculation and even some healthy paranoia are useful tools. But we need to stay within the bounds of verifiable fact, lest we become the very conspiracy-mongering caricatures painted by the mainstream media.
>
> Our attitudes toward the form or forms which extraterrestrial intelligence will take are painfully narrow. This is exciting, intellectual territory, and too many of us have allowed ourselves to be told what to expect by an academically palatable elite. I think some of the objects in the Cydonia region of Mars are probably artificial. And I think the only way this controversy will end is to send a manned mission. The features under investigation are extremely old and warrant on-site archaeological analysis.[21]
>
> We've learned, painfully, that images from orbiting satellites won't answer the fundamental questions raised by the artificiality hypothesis. We don't have all of the answers, but the answers are within our reach. We can continue parroting the 'answers' offered by self-proclaimed skeptics, or we can proceed with objectivity, caution, and the knowledge that reality is seldom as abiding as we'd prefer. The Face on Mars is not dead. I'll make an even bolder statement: If we haven't conclusively established the presence of life on Mars within the next decade it won't be because it's not there; it will simply indicate that we're not trying hard enough.[22]

The importance of Mac Tonnies's words cannot be understated. Here was a man who was willing to study the Face-on-Mars controversy with a refreshingly unbiased mind that was not wrapped up in belief systems. His observations to the effect that in-depth analysis of some of the more striking features of the face suggested the presence of a nostril and an eye—containing what looked like an iris, no less—add much weight to the theory that the mighty structure is indeed artificial. Not only that: Tonnies made a highly important statement when he noted that the eerily human-like appearance of the face, and the Pyramid-style formations seen nearby, may collectively be suggestive of an ancient, direct link between the nonhuman creators of the face and our very own, long-forgotten, fog-shrouded origins as a species.

In other words, when it comes to the cosmic conundrum that is the Face on Mars, Mac Tonnies left us with a wealth of ideas to muse upon, theories to debate, and hard-to-deny evidence of a once-thriving civilization that, eons of years ago on a faraway world, rose and fell while we were still little more than brutish, savage animals.

Chapter 10
UFOs Down in Bolivia

The latter part of the 1970s saw NASA implicated in the issue of crashed UFOs to a degree that was previously unparalleled (to the best of our knowledge, that is). Indeed, throughout 1978 and 1979, it seems that the activities of NASA and those of crashed UFOs went together practically hand in glove. A highly significant event that falls squarely into this particular category, and which continues to remain shrouded in mystery and obfuscation to this day, occurred in the nation of Bolivia in early May 1978, and is supported by an intriguing and varied body of official documentation that has surfaced from, among other agencies, the CIA, the Department of State, and NASA itself.

In May 1978, a U.S.-based research group called Citizens Against UFO Secrecy (CAUS) began to dig deeply into a sensational story reported by United Press International (UPI) suggesting that an extraterrestrial space vehicle had plunged to Earth somewhere in Bolivia. Not only that: NASA was reportedly deeply implicated in the investigation of the mysterious affair.

Two Days after the UPI story broke, CAUS contacted NASA as part of an attempt to secure an on-the-record comment, and, hopefully, to gain a better and clearer understanding of what had actually taken place. Interestingly, as CAUS duly noted, there seemed to have been a conveniently sudden and widespread attack of bird flu in the NASA Public Affairs Office, as spokesperson after spokesperson was said to be absent and home ill, and thus unavailable for comment. Finally, however, one Debbie Rahn, who was then an assistant to a NASA Public Affairs officer named Ken Morris, and who had seemingly not fallen from the bug sweeping NASA, provided some welcome assistance on the matter.

Rahn told CAUS that from what NASA's Public Affairs Office had been able to ascertain thus far, NASA most certainly had not dispatched anyone to Bolivia to investigate anything at all. Moreover, Rahn asserted firmly and confidently that the UPI report was absolutely false. She was, however, willing to refer CAUS to a certain Colonel Robert Eddington, who was working in the Department of State in 1978. Eddington went on the record to CAUS, saying:

> [NASA] have had numerous inquiries and immediately contacted us. We have received communications from our people who have also seen newspaper accounts. What we do not have is any kind of firsthand information that, in fact, the object does exist.[1]

Eddington was willing to admit, however, that he had received a number of reports of the unknown object being oval in shape and "solid," and speculated that it was very possibly some form of "near spherical liquid oxygen/hydrogen tank from a booster." Nevertheless, he conceded that "...four meters is a big tank." Notably, Eddington also revealed to CAUS that his office was part of the Bureau of Oceans, International Environmental and Scientific Affairs, and "kept track of launches."[2]

Although those attached to NASA who were questioned by CAUS claimed no personal or direct knowledge of anything truly substantial with respect to the Bolivian case, a batch of official documentation on the affair that has since surfaced into the public domain clearly demonstrates that something very strange was afoot in Bolivia—and that NASA *was* being kept informed of each and every development in the matter.

The genesis of the *X-Files*-style affair appeared to be in a U.S. Department of State telegram transmitted from the American Embassy in La Paz, Bolivia, to the U.S. Secretary of State, in Washington, D.C., on May 15, 1978. In 2009, a copy of this telegram was made publicly available by NASA, clearly demonstrating that the space agency had a deep awareness of the case from its beginnings.

Captioned "Report of Fallen Space Object," the document revealed that American authorities had taken notice of the fact that a number of Bolivian newspapers had gotten hold of the basic details of the story, and were specifically referring to an "unidentified object" having "recently fell [sic] from the sky." Further, files obtained from NASA show the Bolivian media was saying much more than that: A specific rumor was going around Bolivia that the aerial object had been found somewhere relatively close to the Bolivian city of Bermejo, that it was somewhat egg-shaped, that it appeared to be constructed out of an unknown type of metal, and that it was around 13 feet in diameter. NASA was further informed that senior elements within the Bolivian Air Force planned to carry out an in-depth, ambitious study of the device as part of an effort to ascertain its point of origin. Perhaps most significant of all, the Department of State advised NASA that the "general region has had more than its share of reports of UFOs the past week. Request a reply ASAP."[3]

The Department of State was not the only branch of officialdom that took a secret interest in the Bolivian matter, as a CIA report—also of May 15, 1978—makes abundantly clear. Significantly, copies of the CIA documentation that follow were also made publicly available via NASA in 2009. In other words, the space agency was closely following the CIA's findings on the nearly unique affair, as well as monitoring the Department of State's work, at the time of its occurrence:

> Many people in this part of the country claim they saw an object which resembled a soccer ball falling behind the mountains on the Argentine-Bolivian border, causing an explosion that shook the earth. This took place on May 6.[4]

Echoing the words of the Department of State, the CIA told NASA that "around that time some people in San Luis and Mendoza provinces reported seeing a flying saucer squadron flying in formation." The CIA also stated, with a high degree of confidence, that the "artificial satellite," as it was then being described, had impacted on Bolivia's Taire Mountain and had already been found by Bolivian military personnel. The CIA added, in concerned tones, something that clearly demonstrated the serious nature of what was afoot: "The same sources said that the area where the artificial satellite fell has been declared an emergency zone by the Bolivian government."[5]

Yet another CIA report also located within the archives of NASA referenced the crash and added important data to that already in hand. Dated May 16, 1978, and titled "Reports Conflict on Details of Fallen Object," it highlighted the fact that the CIA was doing its utmost to get a full understanding of the nature and gravity of the situation, and—despite what NASA was saying to the contrary—had placed NASA deep in the heart of the controversy:

> We have received another phone call from our audience requesting confirmation of reports that an unidentified object fell on Bolivian territory near the Argentine border. We can only say that the Argentine and Uruguayan radio stations are reporting on this even more frequently, saying that Bolivian authorities have urgently requested assistance from the U.S. National Aeronautics and Space Administration in order to determine the nature of that which crashed on a hill in Bolivian territory.[6]

The CIA also said that only a few minutes prior to the preparation of its latest bulletin on the incident, *Radio El Espectador*, of Montevideo, had made an announcement to the effect that there was a degree of uncertainty and conflict within Bolivia as to the legitimacy of some of the stories surrounding the crash. From reputable contacts within Argentina, however, the CIA uncovered information—which it quickly forwarded on to NASA—indicating that:

> ...the border with Bolivia had been closed but that it might soon be reopened. They also reported that an unidentified object had fallen on Bolivian soil near the Argentine border and that local Bolivian authorities had requested aid from the central government, which, in turn, had sought assistance from the U.S. National Aeronautics and Space Administration to investigate the case.[7]

Perhaps most significant of all was a story making the rounds in both Bolivia and Argentina, positing that not only were Bolivian authorities extremely interested in uncovering more about the nature and origin of the unknown device, but that, in addition:

> Local authorities, for security reasons, had cordoned off 200 km around the spot where the object fell. The object is said to be a mechanical device with a diameter of almost 4 meters which has already been brought to Tarija. There is interest in determining the accuracy of these reports which have spread quickly throughout the continent, particularly in Bolivia and its neighboring countries.[8]

The CIA asked NASA:

> Is it a satellite, a meteorite, or a false alarm?[9]

On May 18, 1978, the U.S. Embassy in La Paz again forwarded a telegram to the Secretary of State, Washington, D.C. classified at Secret level. The telegram was also shared with NASA and again reflected the desire of numerous official U.S. intelligence agencies to ascertain the truth and the full picture of what was occurring in Bolivia, and the nature of what had crashed and reportedly been recovered.

Six days later, a communication was transmitted from the U.S. Defense Attaché Office in La Paz to a variety of US military and government agencies, including NASA, the North American Air Defense Command (NORAD), the U.S. Air Force, and the Department of State. Its contents make for intriguing reading:

> This office has tried to verify the stories put forth in the local press. The Chief of Staff of the Bolivian Air Force told DATT/AIRA this date that planes from the BAF have flown over the area where the object was supposed to have landed and in their search they drew a blank. Additionally, DATT/AIRA talked about this date with the Commander of the Bolivian army and he informed DATT that the army's search party directed to go into the area to find the object had found nothing. The army has concluded that there may or not [sic] be an object, but to date nothing has been found.[10]

TOP SECRET

So what, exactly, did occur on that day in May 1978? Whereas the available U.S. government records that have surfaced via NASA certainly point toward the probability that something very out of the ordinary occurred, which was of clear and demonstrable interest to whole swathes of the U.S. intelligence community, somewhat frustratingly those same documents also raise far more intriguing questions than they provide hard, definitive answers—as, unfortunately, is often the case when it comes to trying to unravel and understand the nature of official secrecy and its relationship to the UFO mystery.

Consider the facts, such as they were and still are: The CIA's report of May 15, 1978, to NASA clearly stated that the object had fallen to earth on Taire Mountain, Bolivia, and had already been located by Bolivian authorities. Furthermore, on the following day, the CIA specifically learned that the object had been taken to Tarija. In stark contrast, however, the Bolivian Army and Air Force both advised and assured the U.S. Defense Attaché Office that their search for the mystery object had drawn a complete blank and nothing was found. Were both NASA and the CIA misinformed by Bolivian authorities? Was the Bolivian military intent on keeping American agencies, including NASA, in the dark? Regardless of the answers to those particularly controversy-filled questions, the story most certainly does not end there.

By June of 1978, CAUS had spoken with the respected UFO writer and researcher Bob Pratt, who also happened to be the *National Inquirer*'s UFO authority, and the person who had been dispatched to Bolivia specifically to cover the story in question while it was still relatively fresh and able to be investigated to a fairly significant degree. While he was on location, Pratt had the very good fortune to speak with a number of Bolivian witnesses who reported seeing the UFO perform a series of starling maneuvers in the sky before it exploded and finally fell to the ground. According to Pratt's sources, there were actually two explosions: The first was a massive one reportedly heard at a distance of more than 80 miles; the second was a mere rumbling in comparison. Pratt also divulged that he had actually had the opportunity to fly over the suspected crash site and had been able to identify what gave all the appearances of being a very recent landslide in which the rocks showed clear signs of significant damage. Pratt was evidently ready to write a story for the newspaper to the effect that a UFO really had crashed in Bolivia, but by June 19, an editor at the *Inquirer* decided not to run with it.

Two months later, further data pertaining to the mysterious crash had been uncovered by CAUS:

> From what CAUS can determine...an expedition of Bolivian army soldiers and scientists...returned from the suspected impact area on May 21. CAUS has reliable information from an American source that this expedition did not get to Cerro Bravo [Bravo Mountain], the suspected crash site, because the slopes were too steep to negotiate.[11]

Then, after the first expedition returned, a Bolivian astronomer, who had been part of a team that flew over Cerro Bravo in a Brazilian Air Force plane, noted a rockslide that convinced him something had struck the side of the mountain, causing the slide. Quite possibly, this was the very same significant rock disturbance to which Bob Pratt was a firsthand witness.

On May 23, a trio of Brazilian military personnel (on horseback, no less) embarked upon a quest to determine the truth behind what had occurred, and finally reached the area in question around 48 hours later. And although they found no remains of any sort of vehicle—alien or otherwise—they were able to confirm the presence of an approximately 300-foot-long gouge, which strongly suggested something unusual had slammed into the ground with considerable force.

UFO researchers Lawrence Fawcett and Barry J. Greenwood reported, with much justification:

> One of the very frustrating things in attempting to obtain documentation for incidents like the Bolivian crash is the extreme reluctance of government authorities to want to give a full accounting of anything.[13]

Nevertheless, on a positive note, there is the fact that other sources have been able to offer at least a modicum of additional insight into this curious affair.

For example, in June 1979, a former U.S. intelligence officer, Leonard H. Stringfield, learned of some potentially highly inflammatory data on the case from an Argentinean investigator named Nicholas Ojeda. Ojeda said to Stringfield at the time in question that a rumor was in circulation concerning a "group of investigators who vanished mysteriously in the area. I really think something big happened in Salta. NASA investigated, but there was no news of it. I have to tell you that in La Paz, Bolivia, a huge Hercules C-130 [aircraft] carried 'something' from the area where the UFO crashed."[13]

In addition, Stringfield's careful research led to a notable disclosure from a CIA source known to the UFO investigator Bob Barry, who confirmed that the C-130 flight did take place, and that he, the informant from the CIA, was aboard that very aircraft. A terse statement of *no comment* was the only reply that Barry succeeded in receiving from the CIA operative when the issue of the specific nature of the aircraft's mysterious cargo was tactfully raised.

To this day, more than 30 years later, the Bolivian mystery of May 1978—and the full extent of NASA's involvement in the matter—remains precisely that: a mystery. The story was not quite over, however. One year later, in 1979, NASA was still monitoring unusual aerial activity deep in the heart of Bolivia.

CLASSIFIED

After having filed a series of routine Freedom of Information Act requests with a variety of U.S. government, military, and intelligence agencies to determine if additional data existed on the Bolivian incident of 1978 that so deeply interested NASA, I received a batch of reports from NASA that originated with the Defense Intelligence Agency that showed in the following year, 1979 (and specifically in the month of August), a number of strange objects had been found on farmland in the Santa Cruz area of Bolivia. Once again, the files in question had been circulated among the higher echelons of NASA.

Although there does not appear to be a direct link between later events and the incident of 1978 that so intrigued and concerned NASA and the CIA, the evidence of yet more anomalous aerial activity in Bolivia—all of which greatly interested NASA—is highly compelling and worthy of both comment and scrutiny. The files begin with the following brief summary-report from the Defense Intelligence Agency to NASA:

> On late afternoon [August 8, 1979] the Embassy here received information that a strange object had been found on a farm near Santa Cruz, Bolivia. Source stated that the object was about 70 centimeters in diameter and two meters in circumference with a hole in one side and a metal skin covering of approx one-half-inch thickness. Later the object was described as "about three times the size of a basketball."[14]

A second report that was found in NASA's archives—one that was filed shortly after the events described previously occurred—shows that American authorities were taking a considerable interest in the situation, and also demonstrates that a second object had come to ground on the same day, but this time on farmland located some 125 miles north of the city of Santa Cruz. Somewhat mysteriously, one Jan Saaveedra, on whose property the object had apparently not crashed but actually landed smoothly, said, not only had he "heard a loud whistling sound and saw [sic] a fireball followed by an explosion," but also, on the night that directly followed the incident he had seen a "silent aircraft that had three lights," which was specifically "flying over the explosion area."[15]

CONFIDENTIAL

A perusal of the files from NASA reveals that major steps were taken to preserve the recovered materials and to chronicle the events in question. NASA records demonstrate that numerous color photographs were taken of the object, and movie footage of the landing site and its surroundings was secured.

Questions, of course, abound, and remain unanswered: From where did the spheres originate? What was the nature and origin of the silent aircraft that flew over the crash site displaying three lights? And how had at least one of the mysterious spheres apparently negotiated a completely smooth landing, rather than a crash? Perhaps the photographs and the movie footage in the possession of the Defense Intelligence Agency (and all copied to NASA) would supply the much-needed answers to those puzzling questions. Interestingly, although confirmation has been obtained that NASA did receive its own copies of the film footage and all of the photographs, the priceless material appears to have summarily vanished from the archives of both NASA and the Defense Intelligence Agency—a fact that echoes certain aspects of the Kecksburg, Pennsylvania, affair of 1965, you will recall.

TOP SECRET

But still this story is not quite at an end. Throughout the years, I have become very accustomed to receiving through the mail all manner of envelopes and packages adorned with official emblems and stamps, after filing Freedom of Information Act requests with U.S. government agencies. On filing requests specifically for data on the Bolivian crashes of 1978 and 1979, however, each and every envelope I received was adorned with a small emblem with which I was not at all familiar.

Closer scrutiny, and a bit of research, revealed that the emblem was, in fact, that of a little known body within the British Ministry of Defense known as DIALL—the Defense Intelligence Agency Liaison Office at Whitehall, London, England.

Although I am British by birth, I was living in the United States at the time that the documentation from NASA surfaced, so I wondered, what was so important about the Bolivian events that it required such a seemingly high-level Anglo-American liaison? Why were the replies to all my FOIA requests being routed through the British Ministry of Defense in London, England? Was it possible that, as well as NASA and the DIA, British authorities were also somehow implicated in the Bolivian events of 1978–1979? The possibility is one that we would be very wise to consider, for reasons that may be as mystifying, and as mysterious, as the crashes themselves. And we are still not quite done with the connection between certain elements of the British Government and NASA.

CLASSIFIED

From a former employee of the British government's Home Office comes a truly fascinating story that clearly demonstrates the way in which senior personnel from NASA have secretly liaised with colleagues in the British Isles on the subject of crashed UFOs—possibly for decades. It was in 1996 that Matthew Williams, formerly a special investigator with the British government's Customs & Excise Agency, provided me with a copy of a British government Home Office–originated document titled "Satellite Accidents with Radiation Hazards" that had been written in 1979, and that had been circulated to every chief of police, every chief fire officer, and every county council in England and Wales, with similar documentation circulated to official bodies throughout Scotland as well.

An examination of the file at issue makes it very clear that the Home Office's decision to circulate the classified document on such a large scale was, ostensibly at least, because of an event that had occurred 12 months prior, as the following extract from the file reveals:

> Following the descent of a nuclear-powered satellite in Canada on 24 January 1978, consideration has been given to contingency arrangements for dealing with the possibility of a similar incident in the United Kingdom.

The Home Office added that the possibility of a nuclear-powered satellite crashing somewhere within the British Isles was considered to be fairly "remote." The Home Office was, however, very careful and wise to add that "the special considerations that affect the use of nuclear materials and the safety standards applied to them make it prudent to devise plans to deal with such an incident."[17]

Although, at least at first glance, this extract from the document does indeed seem to focus its attention solely upon the threat posed by decaying space satellites of a distinctly terrestrial nature, Peter Jeffries, who worked for the Home Office in the 1970s, strongly claimed otherwise. Jeffries stated firmly that the document was prepared "according to the guidelines of some of the most senior people in government and with the assent of NASA," which had an ulterior motive in mind.[18] Jeffries explained what he meant by this:

> Yes, the document was drafted to deal with the possibility of a satellite—like Skylab, for example—hitting Britain. But this was in many ways the umbrella, the cover story. This was a project to also allow for us and NASA to address the recovery of crashed UFOs—or downed UFOs—but to be able to keep it covered up by hiding it all behind the cover story of crashed-satellite incidents. Get inside the Home Office's crashed-satellite regulations, and you'll find buried in there its history of crashed UFOs, and its time spent working with NASA on the same subject. I can say no more.[19]

When questioned on this matter, representatives of the Home Office denied it had even the remotest basis in reality—which is, of course, precisely what one might expect them to say if Peter Jeffries was speaking truthfully.

One other important issue to bear in mind is this: There is an acceptance on the part of many people (probably an unconscious one) that any visiting extraterrestrial entities to our world are going to be nearly omnipotent in nature. As this chapter's revelations and documentation concerning NASA and crashed UFOs would seem to indicate, however, neither our alien visitors nor their craft are apparently infallible. They are as prone to mistakes, to errors in judgment, and perhaps even to mechanical, scientific, and technical flaws as are we. In other words, if the UFO intelligences do one day prove to be hostile to the human race, we do stand at least a chance of taking them on in a combat-driven situation.

Chapter 11
Space Shuttle Sabotage

There can be no doubt that NASA's introduction of the space shuttle, which underwent initial tests in 1981 before commencing with regular flights into space in the following year, revolutionized both space travel and the agency's leading role in the domain of otherworldly activity. But, major and now historic advances in space exploration aside, there is distinct high strangeness and cosmic conspiracy attached to the space shuttle program, most of which revolves around the catastrophic losses of the shuttles *Challenger* in 1986 and *Columbia* in 2003. The earliest such example involves none other than the renowned Carl Sagan.

Born on November 9, 1934, in Brooklyn, New York, Sagan was an undoubted prime mover in a number of highly significant and important NASA operations and projects designed to explore the planets of our solar system. And, for his many achievements and contributions to the world of space exploration, Sagan was justifiably honored and rewarded. He became the recipient of NASA medals for Exceptional Scientific Achievement and for Distinguished Public Service, as well as receiving the NASA Apollo Achievement Award. In other words, not only was Sagan a brilliant thinker and a true visionary, but he was also an integral player in the fast-paced race for outer space.

Perhaps more than anyone else, Sagan, whose 1980 television series *Cosmos* turned him, with astonishing speed, into a celebrity all around the world, managed to successfully fuse mainstream science, science fiction, and entertainment for the eager viewing masses. Indeed, this fact was borne out most graphically in 1998 when Sagan's science-fiction novel *Contact* was transformed into a blockbuster Hollywood movie of the same name starring actors Jodie Foster and Matthew McConaughey.

In 2000, to the complete and utter astonishment of the vast majority of his friends and colleagues, it became publicly known that Sagan had been clandestinely involved in a very unusual and deeply controversial project, at the heart of which was a late-1950s plan to launch nothing less than a fully armed nuclear weapon into space and then explode it on the far side (or, as some people prefer to call it, the *dark* side) of the moon. Within official circles, the secret operation was dubbed with the innocuous and attention-avoiding title of Project A119. Dr. Leonard Reiffel, a physicist responsible for the project in the latter part of the 1950s, revealed the details of the extraordinary plan and explained that, for the most part, the "proposed detonation" was nothing more than an exercise in public relations. In other words, therefore, any possible and theoretical scientific advances that might conceivably have been made as a result of the project, and the planned atomic detonation, were wholly secondary to the psychological effects that the program would undoubtedly have had on the world's population of the 1950s.[1]

The primary, secret goal of the leading lights within the U.S. Air Force was to try and ensure that the explosion would be visible to the people of Earth, and most importantly of all, to the Soviet hierarchy, to the movers and shakers within the Kremlin, and to senior elements of the Russian military. Project A119, therefore, had it ultimately come to meaningful fruition, would have been an exceptional piece of one-upmanship and a clear, forceful demonstration of U.S. military, scientific, and technical superiority in the burgeoning arena of outer space. Needless to say, history has demonstrated that Project A119 was ultimately shelved, due, in part, to the potentially hazardous scenario of launching a nuclear weapon into space and the real risk of it then catastrophically crashing back to Earth, with its deadly, radioactive cargo still intact and exposed to the elements.

This was not the only occasion when high-level secrets of a distinctly outer-space and conspiratorial nature brought Carl Sagan and the government together. In the early 1980s, Sagan became the subject of a secret 33-page FBI file, and related investigation, after he had been involved in a very strange affair, part of which was focused upon

NASA's ultimately ill-fated *Columbia* space shuttle. On November 16, 1983, the FBI's Special Agent in Charge at Cleveland, Ohio contacted the Washington, D.C., offices to report an odd and sinister development in the life and career of Carl Sagan. The Cleveland office ultimately recorded the following in a short memorandum to the director of the FBI:

> On November 15, 1983, Dr. Carl Sagan, Space Sciences Building, Cornell University, Ithaca, New York, received a handwritten envelope postmarked Cleveland, Ohio, November 10, 1983, containing a two-page typewritten letter. Letter comprised of cover page dated November 9, 1983, stating it was an open letter directed to various news publishers and identifying Sagan as an influential person to convince others veracity [sic] of message. Both message and cover letter bear typed name of author as M. Springfield.[2]

The letter to Sagan from the mysterious M. Springfield was very brief in content, but was, without doubt, highly unusual. It read like something straight out of the wildest science-fiction story: "The message is so important that I want you to witness that you have received it before November 22, 1983. You have been chosen because of your standing in the community. I believe you are a person of integrity with the ability to convince others that this message is true," the letter began. From there onward, the communication to Sagan went distinctly downhill.

Springfield summed up the allegedly dire situation in an extremely weird rant that began thus: "An Open Letter to All. Warning! Armageddon is coming!" On November 22, 1983, Springfield told Sagan, terrorists would explode a bomb in either a warehouse or a market that was providing free food to impoverished people in San Salvador, but which, in reality was designed to divert people away from a planned attack on a nearby fuel storage installation. Then, suddenly moving on to a completely unconnected topic, Springfield added that if the next launch of the *Columbia* space shuttle went ahead on schedule, there would be a catastrophic explosion in the rocket, "due to a fuel leak." And worse was to come, claimed Springfield: in May 1985 war would begin in the Middle East, with the only victor being Israel.[3]

Springfield then made another strange prediction:

> The American presidency will be Reagan-Bush-Bush and a democrat in '96 whose birthday is the same year as our first president. He will be our last.[4]

Springfield continued with a whole series of predictions:

- Poland would be free by 1990.
- New York and San Francisco would soon both disappear into the oceans.
- A third World War would begin in 1998.
- The world would be poisoned, the result being that life would be practically extinguished by apocalyptic events.[5]

On receipt of this extremely strange missive, Sagan quickly, and maybe not surprisingly, contacted the FBI, which swung into action, as a formerly Secret FBI document of November 18, 1983, noted:

> A search of indices, State Law Enforcement Computer, and local directories, failed to locate a person identifiable with the author of captioned letter, described as "M. Springfield." Cleveland telephone directory shows a listing for an "M. Springfield" which address is a large apartment complex. A suitable pretext telephone call to that address revealed that "M. Springfield" at that address died in 1972 and widow now resides there. She had no knowledge of a letter sent to a Dr. Sagan.[6]

Additional FBI documentation now in the public domain demonstrates that bureau special agents went to great lengths indeed to try and identify the source of and motive behind the letter. The original letter and envelope were quickly sent to an FBI forensics laboratory for analysis, but in an FBI report on the matter that was concisely entitled "Bomb Threat" it was revealed that unfortunately "no latent prints of value" could be found on either the letter or the envelope, and the forensic analysis was brought to a frustrating close without any form of success having been achieved.[7]

Notably, although Sagan was himself an entirely innocent party in this unusual affair, extremely detailed background checks on Sagan, his contacts, his associates, and his career were secretly made by the FBI with other U.S.-based government and intelligence agencies to determine if there was any particular reason why he, personally, was singled out to be the one and only recipient of the letter: "No derogatory information on Subject was found after consultation with [Deleted], [Deleted], [Deleted], and [Deleted]," was the FBI's official final word on Carl Sagan and the elusive M. Springfield.[8] Bizarre conspiracies concerning NASA and its space shuttle fleet only continued to surface and spread, however, and with very disturbing frequency.

TOP SECRET

At shortly before noon on January 28, 1986, the space shuttle *Challenger* suffered a disaster a little more than a minute after takeoff from the Kennedy Space Center, which led to the tragic deaths of its seven crewmembers: Commander Dick Scobee, pilot Michael J. Smith, mission-specialists Ronald McNair, Judith Resnik, and Ellison Onizuka, and payload-specialists Gregory Jarvis and Christa McAuliffe. Published within the pages of the 1986 "Report of the Presidential Commission on the Space Shuttle Challenger Accident," the conclusions and findings of NASA, as well as those of additional investigative agencies that assisted NASA in the complex investigation, were that the destruction of *Challenger* had been caused by "a failure in the joint between the two lower segments of the right Solid Rocket Motor. The specific failure was the destruction of the seals that are intended to prevent hot gases from leaking through the joint during the propellant burn of the rocket motor."[9]

As a result of this truly shocking loss of life, and due to the fact that one of NASA's worst nightmares had now finally come all too true, for almost three years the entire space shuttle program was placed on hold. NASA had suffered a setback of previously unparalleled proportions. But, in direct contrast, and while the space shuttles were completely grounded, many conspiracy theories relative to NASA and the destruction of *Challenger* began to blossom, bloom, and take to the skies.

In mid-November 2003, a man named Martin Black, a self-described private eye with an obsessive interest in conspiracy theories, made the wholly unsupported and deeply controversial claim that NASA had itself sabotaged the *Challenger* mission because, on his return to Earth, *Challenger* crew member Ellis Onizuka planned to reveal to the world his personal, "firsthand knowledge" of the U.S. military's recovery of alien bodies at Roswell, New Mexico, in the summer of 1947; they were bodies rumored to be preserved in cryogenic states at an underground installation somewhere deep in the heart of Nevada—possibly at the infamous and ultra-secret Area 51 site.[10]

Of course, even if such a fantastic and outlandish tale possessed the remotest basis in reality, one major question has to be answered: Why on earth would government assassins go to the extent of blowing up the *Challenger* space shuttle to achieve their dark and disturbing aim when, presumably, any skilled and capable hit man could have taken out Onizuka in a far more down-to-earth and much less visible fashion, such as in a car accident or a faked suicide? Black had no logical or reasonable answer to that highly important and deeply relevant question. Very curiously, however, there actually is a real link between NASA's space shuttle astronaut Onizuka and tales of the U.S. military having in its possession a number of extraterrestrial corpses recovered from one or more UFO crashes.

CLASSIFIED

In 1989, former U.S. military intelligence officer Leonard H. Stringfield revealed that, four years earlier, he had spoken with Chris Coffey, a resident of Cincinnati, who had been a close friend to Onizuka. Coffey confided in Stringfield that Onizuka had a profound tale to tell of astounding proportions that was focused directly upon the controversial issue of crashed UFOs and dead aliens held by elements of the American military.

Based on what he had been told by Coffey, Stringfield revealed that Onizuka "kept an open mind on the subject and added that his curiosity was aroused when he and a select group of Air Force pilots, at McClelland AFB [Air Force Base] in 1973, were shown a black-and-white movie film featuring 'alien bodies on a slab.' In his state of shock, he said he remembered saying aloud, 'Oh, my God!' Chris [Coffey], knowing my work in C/R [UFO Crash/Retrievals], had arranged for me to meet Onizuka to discuss UFOs after his scheduled flight on the space shuttle *Challenger*. As it turned out, fate intervened when the shuttle exploded."[11]

And on this same matter, UFO authority Richard Dolan, who has penned a number of groundbreaking books on the UFO phenomenon and its national security considerations and implications, commented that if such film footage is indeed the genuine article, then there might very well be some latent value in revealing it to military personnel. Dolan speculated: "Say that your job is to 'manage' this information" and you have an awareness of the fact "that certain individuals in the military are more likely than others to encounter the reality of UFOs in the course of their career. Showing such a movie could be one method by which to screen potential candidates for special ET-related assignments."[12]

Dolan wondered, when exposed to such mind-expanding film:

- ✔ How might the relevant people cope with, and assimilate into their worldview, such astounding material?
- ✔ Could they even begin to cope with it?
- ✔ Might shock and mental collapse be the only, inevitable results?
- ✔ From a security perspective, could the chosen individuals be capable of keeping such a secret, and could they completely guarantee not to discuss it outside of confidential, official circles?

Dolan's questions are certainly all relevant and thought-provoking. Thus far, however, they remain stubbornly unanswered by officialdom.

There is one other slightly eerie issue on this matter that is also worthy of some note. Born in 1943, the late well-known country singer John Denver had a very deep passion and enthusiasm for the mysterious realm of outer space. He even chose to take, and, notably, passed with flying colors NASA's rigorous physical and mental evaluation to determine if he was fit enough to cope with the many rigors of a potentially perilous journey outside of our Earth. As a result of his success, plans were officially made by NASA for Denver to fly on the ill-fated *Challenger* flight that killed astronaut Onizuka and his colleagues. Fortunately for Denver, he was unable to take to the skies on that tragic day. Had he done so, and had *Challenger* not exploded right after it took to the skies, maybe Onizuka and Denver would have discussed the subjects of UFOs, dead aliens, and the Roswell affair, because the country star was born in the town of Roswell, New Mexico—and into an Air Force family. Denver could not escape disaster in the skies for too long, however. He died on October 12, 1997, when his private plane crashed into the ocean near Pacific Grove, California, an event that the aforementioned Martin Black has also, you may not be surprised to learn, weaved into his complex and controversial theory that elements of NASA secretly sabotaged the January 1986 *Challenger* flight.

CONFIDENTIAL

On January 29, 1986, a document was dispatched to the director of the FBI, William H. Webster, by a special agent who was based at the bureau's Boston, Massachusetts, office, that told an extraordinary and controversial story. Two days before the destruction of the *Challenger* space shuttle, declassified FBI files now reflect, the newsroom at Boston's *Channel 7* television station had received a very worrying call from an unnamed source that had a direct bearing on the space shuttle explosion. It was at 8:35 p.m. on the night at issue, recorded the FBI, when "the caller indicated that he was part of a group of three people who were going to sabotage the shuttle, causing it blow up and kill all aboard."[13]

Astutely realizing that, even if the call might have been nothing more than a hoax or fantasy of very bad taste, it simply could not afford to ignore the matter, the FBI quickly ordered a number of its special agents to descend upon *Channel 7* with the utmost haste, and to speak with the particular person at the station who had taken the call. The FBI agents extensively interviewed staff at the station and were told that the mysterious caller spoke of a series of "horrible, horrible things [that] were going to happen," and that "at least five people are going to be killed" by a secret group that was said to consist of three individuals. The space shuttle *Challenger*, the caller had starkly claimed to the news channel, was about to fly its very last mission.[14]

Despite the problematic fact that the man chose to leave no name, the FBI actually had a very good idea of who he might have been, as is clearly evidenced by the following quote extracted from the FBI's files on this particular matter:

> During briefing of SAC [Special Agent in Charge], ASAC [Assistant Special Agent in Charge], and appropriate supervisory personnel relative to aforementioned and employment of agent personnel, it was recalled that in September of 1985, a walk-in complainant, of questionable mentality, had intimated that he had been responsible for the delay of previous Shuttles, plane crashes, and other catastrophic events.[15]

The FBI quickly set about trying to find the man, and they soon did so: He was found nonchalantly eating dinner at a nearby Frank's Steak House, where he was duly apprehended and arrested. It was noted immediately, and diplomatically, too, by the FBI agents present at the scene, that the man was clearly "not in possession of full faculties." And, as a direct result, he was released into the specific care of an unnamed, local medical unit for a "five-day mental evaluation."[16] Ultimately, as a result of his dazed and confused condition, no prosecution was ever brought against the still-unidentified individual.

It is ironic that although the man in question was

- clearly mentally disturbed to a very significant degree
- had no real intention of ever blowing up *Challenger* or any of NASA's other space shuttles
- had made similar, previous threats and claims to the effect that one of the shuttles was going to be destroyed (none of which had ever come to any sort of fruition)...

On this occasion he got it practically spot-on: shortly after the man's call to the newsroom of *Channel 7*, both the shuttle and its entire crew really were gone. Sometimes, truth actually is stranger than the wildest of all science-fiction scenarios. The final word on this particular matter went to the FBI, an agent of which recorded, somewhat wearily and warily, one strongly suspects:

> It is entirely feasible, and in all probability likely, that [the man] will make similar calls prior to departure of future Space Shuttles.[17]

Currently available FBI files do not reflect any evidence that the man at issue crossed paths with agents of the bureau ever again; however, there were others, just waiting and looming in the wings, with their own dark conspiracy theories pertaining to NASA and the destruction of *Challenger* and its doomed crew.

TOP SECRET

In the week that the FBI's Boston-based office was diligently pursuing a NASA space shuttle conspiracy theory, FBI agents based on the West Coast of the United States were doing the same. An April 18, 1986 FBI document titled "Space Shuttle Challenger, Information Concerning Launch Explosion, Kennedy Space Center, Florida, January 28, 1986" recorded the facts, as brief as they were:

> On January 31, 1986, the FBI Resident Agency in Santa Ana, California, was advised by [source's name deleted] that he believes the Challenger exploded due to its being struck by laser beams fired from either Cuba or an aircraft. [Source] stated that a review of film footage of the explosion revealed brown puffs of smoke coming from the space shuttle just prior to the explosion. He stated leaks from the fuel tanks would produce white smoke, not brown smoke. [Source] said that the brown smoke would be produced each time the craft took a "hit" by the laser beam, and the explosion occurred when the laser beam penetrated the skin of the craft.[18]

After quickly speaking with U.S. Army, Navy, and Air Force specialists that were then working in the growing field of laser-based weaponry, the FBI recorded that the "laser theory is plausible," but was careful to stress that it was considered "not likely."[19] As a result of this particular statement, the matter was dropped—by both the FBI and NASA security personnel, who had been quickly informed of the man's theory when it had reached the attention of the FBI.

CONFIDENTIAL

One month after the *Challenger* disaster occurred, both the Tampa, Florida, and the Dallas, Texas, offices of the FBI found themselves embroiled in a similar wild conspiracy theory. An FBI report of March 1, 1986, that originated with the Tampa office revealed that colleagues in Dallas had been contacted by a man who was extremely reticent about revealing his identity, his home address, and telephone number "for fear of retribution from his employer for using equipment for private interest without authorization," but who had an amazing theory to impart that concerned the events of January 28, 1986.[20]

The caller was willing, somewhat grudgingly, to admit that he was in the employ of a Dallas-based company whose specialty was the enhancement of video via computer, and continued that he had recorded the footage of the *Challenger* explosion from the nightly television news, and had then elected to "run his video on a computerized enhancement machine."[21]

As special agents of the FBI at Dallas listened carefully and intently, the man explained that just before seeing a "puff of smoke" that came prior to the main explosion of the space shuttle, his attention had been drawn to an "indentation on one of the boosters, which was about the size of a basketball." The caller's conclusion, FBI agents at Dallas told their colleagues in Tampa, was that "something hit the rocket booster, subsequently causing the explosion."[22]

In this case, all of the relevant data, which included the notes of the telephone call, and the surrounding official documentation, was forwarded on to NASA by the FBI. A study by NASA personnel of the man's story led them to believe that it was completely without any foundation at all. All talk of the man's theory was quickly dismissed; however, there is nothing contained in the file to suggest he was anything other than a loyal American trying to offer some degree of help to the investigation of the destruction of *Challenger*.

CLASSIFIED

At the same time that both the Dallas, Texas, and Tampa, Florida, offices of the FBI were embroiled in their controversy, the strangest of all conspiracy theories relative to NASA and the space shuttle tragedy of January 1986 was quietly but quickly unfolding at the bureau's Washington field office. The story is a truly odd one, indeed, involving psychic phenomena, the reported complicity of one of the astronauts in the disaster, and the presence of a distinctly shadowy and all-powerful Japanese terrorist group. Interestingly, several extracts from the FBI file on this spectacularly odd affair have been summarily blacked out for reasons specifically relative to the national security of the United States of America.

According to the heavily redacted documentation that the FBI has been willing to release into the public domain, a source of some standing (one who was apparently very well known to Bureau agents) had personally met with a woman—whose name the FBI's censors have been extremely careful to completely delete from the presently available papers on the matter—who "claims to be in contact with certain psychic forces that provide her with higher information on selected subjects. She refers to these forces as 'Source' and when providing information from Source she often speaks in the collective 'we.' [The woman] claimed that she had come to Washington, D.C., to provide information concerning the *Challenger* space shuttle explosion on 1/28/86."[23]

FBI files reflect that a personal, face-to-face meeting with the woman took place on February 24, 1986. With an audiotape recorder on hand, FBI agents sat back in their chairs and listened to a story of truly unusual and surreal proportions. The destruction of *Challenger* had absolutely nothing whatsoever to do with the spacecraft itself, or with any wrongdoings, errors, or mistakes on the part of NASA, the woman confidently informed the several FBI agents who were present at the interview. Rather, she said to one and all, the terrible events of January 28, 1986, were all the work of terrorists. Not only that: two of those individuals whom the woman claimed were involved in the successful operation to destroy *Challenger* were ground workers at the Kennedy Space Center, while the third, outrageously, was said to have been one of the astronauts that had been destined to take part, and subsequently die, in the mission itself.

As for the terrorist group itself, if it existed, then it was truly a unique one. The group was reported to be "fanatical" in nature, possessed of what was specifically described as an "ancestral lineage," and had an overwhelming, admitted hatred of the people of the United States of America and their government. The plan of the terrorists, FBI agents were told, was to bring deep embarrassment and anxiety to NASA, and to the office of the American presidency, and to ensure that NASA's space program suffered a major setback—all of which, with the benefit of hindsight, is precisely what did actually occur, regardless of how much faith one places, or one does not place, in the claims of the woman and her unidentified psychic informants."[24]

True or not, the woman had weaved together an extremely intricate tale, as the FBI carefully noted in its detailed files prepared directly after the interview was completed:

> The explosion was effected by a device placed inside the external fuel tank of the shuttle. An individual whose description seems to match that of an engineer or technician placed this charge. The charge was triggered by a second saboteur using a hand-held transmitter while standing in the crowds watching the shuttle lift-off. The individual matches the description of a guard or security person. The astronaut saboteur chose to die in the explosion as a sort of ritual death or "cleansing."[25]

Very noteworthy is the fact that, not only did the woman provide the FBI with data on the nature of the work that two of the saboteurs were engaged in at NASA, but also that the physical descriptions of the pair that were also given to agents of the FBI were "probably complete enough to pinpoint the individuals." If, however, the FBI was indeed able to take this aspect of the investigation to another level—perhaps one of a much higher secrecy classification—then it is most certainly not reflected within the pages of the presently available files on this particular affair.[26]

As for why, exactly, the woman chose to come forward and tell her unusual tale? It was primarily, she advised the FBI, because she had major concerns that the United States faced being bested in the domain of outer space by potentially hostile and aggressive nations if NASA's ambitious programs were placed on hold, or, even worse still, were outright canceled as a result of a tragic event that was, she fully believed and accepted, essentially nothing to do with any wrongdoings on the part of NASA. In other words, for NASA to place its Space Shuttle program on hold for reasons that were not the fault of NASA in the first place, was, said the woman, a major and potentially disastrous mistake.

In essence, that is the extent of the declassified material on this particularly unusual, and probably even unique, space shuttle–based conspiracy theory, but two things in particular are worth noting:

1. The FBI has openly admitted to withholding in their entirety, and in the name of U.S. national security, no less than 26 pages of secret documentation on this specific saga.
2. The reference to the NASA astronaut of "oriental heritage" who died in the explosion can only have been a reference to Ellis Onizuka, who, as has already been carefully noted, has been linked with yet another conspiracy theory tied to the destruction of *Challenger*: namely, that relative to stories of alien bodies said to have been found at Roswell, New Mexico, in 1947, and secret movie footage of such, secretly claimed to be in the possession of elements of the U.S. government and/or the military.

Unless, or even until, the FBI elects to one day release into the public domain the remaining, still classified data and pages on this highly unusual affair (as well as the audiotape recording of the FBI's interview with the unknown woman), we are unlikely to ever know with complete certainty whom the FBI's mysterious source was, or the precise nature of the "certain psychic forces" with whom she claimed enigmatic and ethereal ESP-style contact.[27]

It is worth noting at this point that, although the FBI's decision to meet with someone who claimed that her knowledge of the nature of the space shuttle disaster was derived from psychic phenomena, or intelligences, might sound unusual, and may make many people wonder why the FBI even chose to speak with her in the first place, the fact is that certain elements of the American military and intelligence community have, for decades, investigated psychic phenomena, ESP, out-of-body experiences, and even Ouija boards to determine if they might play meaningful roles in secret, intelligence gathering operations.

Under the terms of the Freedom of Information Act, tens of thousands of pages of formerly classified documents have now surfaced on such paranormal matters from the archives of the Army, the Air Force, the CIA, the Defense Intelligence Agency, and the FBI. Sometimes, those same files clearly demonstrate that when a conventional approach to seeking important data appears not to be successfully working, very often the only remaining option is to follow a path of distinctly unconventional proportions—which, it appears, is precisely what the FBI chose to do in 1986 while deeply embroiled in its nationwide investigation of the destruction of *Challenger*.

The explosion of the space shuttle Columbia in 2003 sparked off a flurry of conspiracy theories.

CONFIDENTIAL

Wild conspiracy theories of a religious nature reigned supreme when, on February 1, 2003, NASA's space shuttle *Columbia* disintegrated over the state of Texas during reentry into the Earth's atmosphere, with the inevitable loss of all her crew. Despite some initial conspiracy theories about the shuttle being carefully targeted for destruction by foreign terrorists (which surfaced directly after a news announcement that the *Columbia* explosion had occurred over Palestine, Texas, and that it was carrying the first-ever Israeli astronaut: Ilan Roman), NASA, and indeed the government as a whole, summarily and rapidly denied the notion that such theories had any basis in reality, and stated that they weren't even worthy of comment. Gordon Johndroe, of the United States Department of Homeland Security, appeared on television, confidently assuring the concerned American populace that there was absolutely no information in hand to suggest that terrorists had played any sort of role, large or small, in the *Columbia* disaster.

This assurance did not, however, go any way toward satisfying those who saw the handiwork of extremists—whether Muslim or Israeli—firmly in evidence with respect to the destruction of *Columbia*. And, for devoted believers in such matters, there were even those who saw the mighty power of God himself at work in the incident. That the space shuttle tragedy had occurred over the town of Palestine, Texas, for example, stimulated graphic imagery of the Palestinian struggle against Israel's occupation of the West Bank. In addition, Ilan Roman was not only Israel's first astronaut, but he was also one of the country's most renowned of all its military heroes: he had served as a colonel in the Israeli Air Force and had piloted a plane that bombed and pulverized Iraq's Osirak nuclear reactor more than two decades previously.

As a result of these facts—these very eerie facts, some commentators and religious zealots were very quick to proclaim—the Saudi-backed daily newspaper *Asharq Al-Awsat* declared that:

> American Palestine has been made famous by the crash of Columbia. Iraq will be happy about the death of the Israeli astronaut who bombed its nuclear reactor.[28]

In a similar fashion, Hezbollah leader Sheikh Hassan Nasrallah said, approximately 24 hours after the destruction of the space shuttle that the catastrophic event was nothing less than a graphic message for the entire human race. It was, he added in deliberate tones, "a message to those who thought in the past few years that America was a god that couldn't be defeated or defied." Nasrallah added that the people and government of America could only watch in awe, "unable to do a thing, as its space shuttle blew up and plummeted to earth. America, like it or not, gave in to the will of God Almighty."[29]

And, all around the world, many of a religious nature and persuasion nodded both gravely and knowingly, despite NASA's very best attempts to place the tragedy of the *Columbia* space shuttle in an overwhelmingly down-to-earth context. NASA's current plan is to retire its fleet of space shuttles in the very near future. Until that day finally comes, however, and for as long as these impressive and largely successful spacecraft continue to fly, no doubt the conspiracy theories that surround them will as well.

Chapter 12

Alien Abductions and NASA

CONFIDENTIAL

Any meaningful, concerted attempt to try and accurately determine when the first so-called alien abduction of a human being by purported extraterrestrial entities took place is inevitably going to be a very difficult task indeed. Most researchers and students of the UFO phenomenon, however, would probably concede that the phenomenon that has today become popularly known as "alien abduction" was relatively unknown until sometime after September 19, 1961. It was on that night that Betty and Barney Hill, a married couple from New Hampshire, were driving home from vacationing in Canada when they were allegedly subjected to a terrifying otherworldly experience.

Despite viewing some form of unusual aerial object in the night sky, and even what appeared to be living entities that could be seen through the craft's portholes, until their arrival back home, the Hills had very little indication that there was actually far more to the encounter than they realized. It later came to light that approximately two hours of time could not be accounted for by Betty and Barney. After some months of deep emotional distress, sleepless nights, and very strange dreams pertaining to encounters with unusual, nonhuman beings, the couple finally sought assistance from Benjamin Simon, then a Boston-based psychiatrist and neurologist.

Carefully subjected to time-regression hypnosis, both Betty and Barney recalled what they believed had actually taken place during that missing 120 minutes (or so). Significantly, the Hills provided very similar accounts of encounters with apparent alien creatures that took the pair onboard some form of alien vehicle and then proceeded to subject them to a rigorous series of physical examinations—a number of which were highly distressing and invasive in nature. The experience of the Hills later became the subject of John Fuller's now classic book, *The Interrupted Journey*, and also a 1975 movie of the same name.

TOP SECRET

By far the most commonly reported creatures said to be present during alien abduction cases are those that have become popularly known as "the Grays." Typically, the Grays are short in stature, around 3 to 4 feet in height, have gray/white skin (hence the name), and their bodies are usually described as being thin to the point of near emaciation. Certainly the most striking and memorable features of the Grays are their heads: they are hairless and overly large in proportion to their bodies, with their ears, nose, and mouth being vestigial at best. Their eyes, on the other hand, are generally described as being black, huge, almond-like in shape, and hypnotic in nature. And since that fateful 1961 night, when Betty and Barney Hill unwittingly added a whole new dimension to the UFO controversy, thousands of people from all across the globe have reported close encounters with the Grays.

One of those that claimed multiple encounters with the Grays is a woman named Sharon, who presently works in a secretarial position at NASA's Kennedy Space Center. She viewed the creatures, in much the same way that P.T. McGavin came to view his alleged alien entities (represented, back in 1973, by the long blond-haired Gavon) as wholly deceptive creatures. But data that she was exposed to on the nature, origin, and intent of the creatures she has encountered are very different from those of McGavin.

Now in her mid-30s, Sharon has had multiple experiences with the short, black-eyed aliens that have become so prevalent within the UFO research arena. In her case, the experiences date back to her early teens, when she recalls having received nighttime visitations from diminutive creatures, which always appeared in groups of three, and would levitate her from her bed while her parent slept, and then transfer her to the interior of a huge, futuristic spacecraft. While onboard the craft—which, Sharon says, was a gigantic vehicle that rested on the sea bed several hundred miles off of the coast of northern California, where she lived at the time—she was always introduced by her alien abductors to a very strange-looking group of young children. There were seven of them, and they all seemed to be a blend of human and alien, in the sense that their bodies and skin color were very "Gray"-like, but their faces were far more human-looking, and they also had fine, white hair on their heads and arms (the Grays are generally characterized as being totally devoid of any and all hair).

Within alien abduction research, such half-human, half-alien figures have become known as *hybrids*. So the controversial theory goes among many researchers of the alien abduction phenomenon: the Grays are on a serious evolutionary decline, and, as a result, they are trying valiantly to stave off extinction by creating a crossbred race that is part human and part Gray. In other words, with the complete destruction of their species not perceived as a viable option, the Grays' only real alternative is to milk us for our DNA and do whatever is necessary to keep at least a part of their civilization and their culture afloat—even if that means merging, quite literally, with the human race.

According to Sharon, the aliens would always encourage her to play with the hybrid children—as if, Sharon speculated, the Grays wished to instill human characteristics and traits in their new offspring, but were unsure how to do so for themselves. The only alternative, Sharon believed, was for people like her (namely the abductees) to do all the work for them, and to help nurture and teach the new crossbred species that one day would replace the original, rapidly declining Grays.

Sharon said that these experiences would typically occur three to four times a year when she was between the ages of 13 and 17. When she reached 18, however, Sharon and her then boyfriend, Douglas, moved to San Diego to live (much to the chagrin of her parents, she admitted), and it was then that things changed dramatically for Sharon. She recalled to me how, on one particular night, when she assumed she was on board the huge, undersea spaceship, Sharon was awakened from the experience by Douglas, who told the completely disoriented Sharon that she had been thrashing about in their bed for at least 15 minutes. And, unsure of precisely what to do, Douglas hoped she would eventually settle down, but was finally forced to wake her up, concerned by the possibility that her stress levels were rising to such a degree that she was about to be physically affected by her deep-sleep nightmare.

Despite their closeness, Sharon had never previously told Douglas about her alien-abduction experiences; however, after she had been awakened midway through what was, she perceived, a typical abduction-like scenario, she came to the shocking realization that she had actually been in bed the whole time, but perhaps in some mystifying altered state of mind. On this very same path, Douglas assured Sharon that regardless of what her mind was trying to tell her about the seeming physical reality of the alien abduction experience, she had clearly never exited the bedroom: he had been keeping a careful watch on her for a full quarter of an hour, and, thrashing around the bed aside, she had not left the confines of the bed even once. And so, as a result, Sharon spent the next few hours confiding in Douglas the extraordinary nature of her earlier, cosmic experiences as a young girl.

Thanks to the presence and the quick actions of Douglas, Sharon now began to seriously wonder if all of her experiences had actually followed this very same path. What if, in all those years of apparent alien encounters, she had always been in her bed, in some weird sleep state? Did this mean that the Grays were the product of her own mind and had no actual flesh-and-blood reality? What of the hybrids? Were they, too, brain-borne imagery and nothing more? Or, were the Grays actually not the denizens of outer space, but of a dream-like reality that allowed them to interact with us when we, too, were in a similar dream state?

Whatever the actual truth of the matter, the abductions and the experiences with the hybrid children continued, even though, while Sharon was experiencing these events and encounters, Douglas was able to confirm that on no occasion did she ever leave the perimeters of the bedroom. The most startling event occurred shortly before Sharon's 19th birthday when, once again (from her perspective at least), she was locked into a definitive alien abduction experience deep below the Atlantic Ocean, during which the experience turned very ominous, and Sharon recalled being strapped down to something that vaguely resembled a dentist's chair. The purpose was even worse: Sharon gained the distinct impression that her Gray alien captors were trying to take control of, and capture, her very essence: namely, her soul, which she came to suspect the aliens fed on in the way of terrifying, emotional vampires from the outer edge.

And although Sharon was not, and still is not to this very day, a religious person at all, at the height of the experience, when her terror levels reached near stratospheric proportions, she suddenly cried out and asked for God to save her from these soul-sucking entities from beyond the veil. Suddenly, the group of Grays all appeared unanimously and simultaneously startled, and quickly vanished before Sharon's eyes. In a moment, she found herself wide awake, and in the safety and comfort of her own bed. Not only was this encounter over, but she was never again the subject of any form of alien abduction experience.

Notably, in the same way that P.T. McGavin was interviewed by a representative of NASA when he discussed with colleagues from NASA his 1973 Contactee-type encounter at Hart Canyon, Aztec, New Mexico, so something very similar happened to Sharon when she too, at the age of 28, accepted a position with NASA at the Kennedy Space Center and began talking about her alien-abduction experiences as a teenager. On one particular morning, several days after she had been discussing her encounters with some work colleagues, a call came down to her office from an unidentified male, asking, somewhat bizarrely, if she could "come outside" and "hang around for a few minutes."[1]

Despite a few concerns and a significant degree of understandable puzzlement, Sharon did as she was asked. After a few minutes, a tall, dark-haired woman, probably in her late 30s approached Sharon, introduced herself as someone within NASA who had an interest in stories of alien abduction, and who said she represented a group within the space agency that allegedly kept a close watch on the phenomenon. There was nothing ominous about the meeting, stressed Sharon to me on several occasions. In friendly and relaxed tones, the woman asked Sharon if she would accompany her to her car, where they could talk in privacy. Although it did not occur to Sharon at the time, she later speculated on the possibility that the woman had deliberately chosen her car as a place to talk because she had already surreptitiously set up a hidden tape recorder somewhere in the vehicle, to record each and every word of the conversation.

Whatever the truth of the matter, the conversation was relatively straightforward—at least as straightforward as any conversation on a subject as unconventional as alien abductions could ever hope to be.

Sharon was asked to relate the general nature of her purported meetings with the otherworldly entities, and particularly so with the hybrid children. She was also asked whether or not she believed that asking for the help of the Christian God to end the abductions had any real bearing on the fact that her abduction experiences did indeed come to an end after making such a plea.

In response, Sharon admitted to the woman that, truthfully, she did not know what to think of the whole situation at all, only that it had dominated much of her life as a teenager, but was something from which, aside from discussing it from time to time with friends and those who exhibited an interest in the subject, she had largely moved on. The woman then proceeded to tell Sharon that there were certain people in NASA, and the government as a whole, who knew that the alien abduction phenomenon was not all that it initially appeared to be, and that the "control of our souls" was believed by NASA to be a central part of the abduction experience."[2]

In addition, Sharon was told, those within NASA that were following the alien-abduction puzzle "know something is going on, but we're not sure what, only that it looks bad for us."[3] And that was it. There were no Men in Black–type warnings made to Sharon not to discuss her experiences with her friends and colleagues at NASA, and she was never again questioned officially, or even unofficially about her abduction experiences as a teenager. It is interesting to note, however, that Sharon's mysterious interviewer was not the only person at the Kennedy Space Center to take an interest in alien abductions and the potential connection between the phenomenon and God, and the human soul.

CLASSIFIED

Joe Jordan, who has a very keen interest in the abduction phenomenon, and who to this day works as a safety specialist at the Kennedy Space Center, is not a part of any clandestine group of the type to which Sharon's contact was apparently attached. He does, however, believe that alien abductions do occur during the sleep state—and the sleep state alone—and suggests that they do not involve the person or persons being transferred to any sort of extraterrestrial spacecraft at all. Moreover, he too believes that a belief in God, and calling upon the help of Jesus and God to break the alien abduction spell, does indeed work. Jordan also believes that the so-called Grays are not aliens at all. Rather, he concludes that they are nothing less than the literal, evil, deceptive minions of Satan himself.

Jordan told me, in 2010:

> UFOs weren't even in my vocabulary until 1992. I was introduced to a local MUFON [Mutual UFO Network] state section director in Orlando, Florida, and became involved and went through the training. And I soon became a state-section director for MUFON. This is when I was introduced to the abduction experience: from some of the people coming to the monthly meetings. These people were claiming to be in contact with the so-called entities responsible for the UFO sightings. So, we decided to look at these reports. I told our investigators: "We can continue to chase our tails by looking at lights-in-the-sky, or we can focus on people who are in the front line." And this seems to be the abductees.[4]

He continued:

> The people don't get taken to a ship; they are physically still in bed. I have a couple of cases where people had the abduction experience while in the presence of a witness who was awake. They didn't go anywhere. They almost went into an unconscious state. This was only for a few minutes, but they came out of it totally exhausted and could talk about what had happened to them, and it would take hours for them to tell it all. But it was just minutes. Like a time-displacement. I'm not sure I would call it a hallucination, maybe more of an apparition, something along the lines of a hologram, but it's still in the mind. These entities can create this experience in our minds and we can

interact with it, and it can leave physical manifestations from the experience. And that's why this is so confusing.[5]

Continuing on this topic, we see that Jordan's views as to why so-called alien abductions are occurring are as controversial as they are undeniably chilling:

The purpose of all this is to deny the reality of Christianity. And, they have probably the best propaganda machine I've ever seen or read about. I believe that's the purpose behind this whole experience. Look at the stories of old of gnomes, fairies, and elves: we wouldn't believe that today. So they come in the emperor's new clothes. And they come in a guise that we will accept. But their purpose is to defeat us and to delude us, so that we will take our focus off the one true God. And if that happens, if the Bible is real, and if the message it shares is real, then the people who succumb to this, their souls are doomed. And I think that's what these entities are trying to do. That's their agenda. The demons know they're doomed to Hell; the Bible teaches that is what will happen to them. But when that time comes, they plan on trying to take as many of God's creation with them as they can. It's a cosmic war.[6]

Do certain figures within NASA, perhaps, secretly share Joe Jordan's views? The experience of Sharon strongly suggests that the answer to that question could very well be: yes, most definitely.

Chapter 13
The Monsters of NASA

Few people have not at least heard of the exploits of the notorious Mothman of Point Pleasant, West Virginia, a bizarre, flying creature that was made famous in the 2002 movie *The Mothman Prophecies*, starring Hollywood crowd-puller Richard Gere, that was based on the book of the same name penned by the late authority on just about everything paranormal and supernatural, John Keel. But, long before the Mothman dared to surface from his strange and ominous lair in the mid 1960s, there was yet another mysterious winged thing that struck terror into the hearts and minds of those who were unfortunate enough to cross its terrible path.

Certainly one of the most bizarre of all the many and varied strange beings that haunt the lore and legend of Texas is that which became known, albeit very briefly, as the Houston Batman. The quintessential encounter with the beast took place during the early morning hours of June 18, 1953. Given the fact that it was a hot and restless night, 23-year-old housewife Hilda Walker, and her neighbors, 14-year-old Judy Meyer and 33-year-old tool plant inspector Howard Phillips, were sitting on the porch of Walker's home, located at 118 East Third Street in the city of Houston.

Walker stated of what happened next:

> ...25 feet away I saw a huge shadow across the lawn. I thought at first it was the magnified reflection of a big moth caught in the nearby street light. Then the shadow seemed to bounce upward into a pecan tree. We all looked up. That's when we saw it.[1]

She went on to describe the entity as being essentially man-like in shape, sporting a pair of bat-style wings, dressed in a black, tight-fitting outfit, and surrounded by an eerie, glowing haze. The trio all confirmed that the monstrous form stood about 6 1/2 feet tall, and also agreed that the strange glow engulfing him was yellow in color. The bat-man vanished when the light slowly faded out, and right about the time that Meyer issued an ear-splitting scream.

Mrs. Walker also recalled the following:

> Immediately afterwards, we heard a loud swoosh over the housetops across the street. It was like the white flash of a torpedo-shaped object.... I've heard so much about flying saucer stories and I thought all those people telling the stories were crazy, but now I don't know what to believe. I may be nuts, but I saw it, whatever it was.... I sat there stupefied. I was amazed.[2]

Meyer added to the newspaper that: "I saw it and nobody can say I didn't."[3]

Phillips, meanwhile, was candid in stating the following:

> I can hardly believe it. But I saw it...we looked across the street and saw a flash of light rise from another tree and take off like a jet.[4]

For her part, Walker reported the incident to local police the following morning.

As a long-time resident of Houston, researcher and author Ken Gerhard made valiant attempts to locate the address on East Third Street where the event took place, and discovered that it is no longer in existence. It has seemingly been overtaken by the expansion of the nearby Interstate 10. Strangely, and perhaps even appropriately, the location has apparently vanished into the void, much like the bat-man did—for a while at least.

Several years after he first heard about the exploits of the bat-man, a close friend of Gerhard told him about the experience of a number of employees at Houston's Bellaire Theater who claimed to have seen a gigantic, helmeted man, crouched down and attempting to hide on the roof of a downtown city building late one night during the 1990s. Perhaps, in view of this latter-day development, we should seriously consider the possibility that the Houston bat-man made a return appearance. Or maybe it never went away at all. Instead, possibly, it has been lurking deep within the shadows of Houston, Texas, for more than half a century, carefully biding its time, and only surfacing after the sun has set, and when overwhelming darkness dominates the sprawling metropolis. There is a very good reason why we should believe that is precisely the case—a reason that implicates NASA.

Desiree Shaw's story of her now deceased father, Frank, an archivist who worked at NASA's Houston, Texas–based Johnson Space Center in the 1980s, is bizarre in the extreme. That does not mean, however, that we should dismiss it as being one of no merit, or one that lacks any meaningful value; only that we should view it with open minds that are receptive to challenging ideas and bold, new paradigms.

It was late one night in 1986, said Shaw, when her father returned from his daily routine at the Johnson Space Center. There was nothing particularly unusual about his lateness, Shaw elaborated, as her father was occasionally required to work late into the night. On this particular night, however, things *were* very different: Her father was clearly deeply distressed and even seemed to be on the verge of lapsing into a full-blown anxiety attack. After Desiree and her mother were finally able to calm him down, Frank Shaw told a wild and extraordinary story: While walking to his car that night, he had seen, to his complete and utter horror, perched on a nearby building, a large man-like figure that was utterly black in color, and that seemed to have a large cape draped across its shoulders and back, with two huge wing-like appendages sticking out of each side of the cape. Looking more bat-like than bird-like, the wings made a cracking noise as they slowly flapped in the strong, howling wind.

The creature, Frank Shaw told his amazed family, had clearly realized it had been seen. Not only that: Shaw gained the very distinct impression that the beast was actually relishing that it had been noticed, and was even seemingly deriving some form of deranged, evil pleasure from the fact that it had struck terror into the heart of Shaw. He could only stand and stare, frozen to the spot in complete, abject fear. The sheer horrific unreality of the situation—seeing a large, dark, gargoyle-like entity looming ominously over him from a rooftop at NASA's Johnson Space Center—finally hit home with full force and Shaw raced for his car, flung open the door, slammed it shut, and then sped off into the darkness. He did not attempt to look back even once.

Not surprisingly, Shaw's family suggested that reporting the encounter to his superiors might not be the wisest move he could make. He agreed—for a short while, anyway. After a few weeks, however, the strange event was still gnawing steadily away at Shaw's mind and nerves, and he eventually confided in his immediate superior, who, to Shaw's great surprise and relief, revealed that this was not the first time such a vile entity had been seen late at night roaming around the more shadowy parts of the Johnson Space Center. In other words, Shaw was not going crazy or hallucinating; the beast was real, and Shaw was not the only witness.

Indeed, a secret file on the matter had reportedly been opened some months earlier; primarily because on one occasion, in the same location where the winged fiend had been seen, the remains of two dead and horrifically mutilated German Shepherds had been found, their bodies drained of significant amounts of blood. As a result of this latest encounter, Shaw found himself grilled intently about his experience by "NASA security people who were flown in from somewhere in Arizona—that much I know," as Desiree Shaw succinctly worded it.[5]

A strong suggestion—albeit not a strict order—was made to Shaw not to discuss the event any further with friends, family, or colleagues, which he most assuredly did not. The story would likely never have surfaced had Desiree Shaw not decided to relate the facts after her father had passed away. The affair is far from over, however. And for the next stage in this odd tale of the monsters of NASA, we have to take a trip to the wilds of Puerto Rico.

TOP SECRET

In September 1959, a groundbreaking paper titled "Searching for Interstellar Communications," written by Cornell University physicists Phillip Morrison and Giuseppe Conconi, was published in the pages of *Nature*. The paper was focused upon the idea of searching for extraterrestrial life via the medium of microwaves. Approximately eight months later, a man named Frank Drake decided to test the theories and ideas of Morrison and Conconi for himself. He did so at the Green Bank National Radio Astronomy Observatory, located in West Virginia—the one U.S. state that, more than any other, just so happened to be definitive Mothman territory. Despite lasting for 150 hours, Drake's search of the heavens for messages from alien intelligences was not even remotely successful. Drake, however, was not to be dissuaded or defeated quite so easily as that.

In October 1961, the very first conference on what became known as the Search for Extraterrestrial Intelligence (SETI) was convened at Green Bank. It was here that Drake unleashed his now-famous and much-championed Drake Equation upon the world, which is an admittedly controversial method for attempting to ascertain the scale of intelligent civilizations that may exist in the known Universe. Since then, SETI has been at the forefront of research into the search for alien life. And NASA, too, has at various times been tangentially involved in SETI-style operations. In 1971, for example, a NASA study group known as Project Cyclops was convened to discuss the feasibility of designing a huge array of radio telescopes to try and detect alien life. At the time, however, the operation was shelved as a result of being far too costly to justify going ahead, and today NASA and SETI are largely separate entities.

However, in the same way that NASA has been linked to SETI, so have sightings of dark cloaked entities and reports of animal mutilations. When Frank Drake elected to make it his life's work to search for alien intelligences, he went down a road that eventually led him to the Arecibo Radio Telescope, which is located on the island of Puerto Rico, and where he eventually rose to the position of director. It was at some point early in his tenure as director in the mid-1960s that a guard at the observatory claimed to have seen a sinister-looking man dressed in a black cloak "walking the narrow trail around the perimeter of the bowl."[6]

The guard was of the opinion that the dark figure was nothing less than a blood-draining—and blood-drinking—vampire. Despite his skepticism, Drake politely accepted the guard's report and agreed to at least take a look at it. About 48 hours later, said Drake, "I really was forced to look into it...because a cow was found dead on a nearby farm, with all the blood drained from its body. The vampire rumor had already spread through the observatory staff, and now the cow incident whipped the fears of many people into a frenzy."[7]

Was it only a coincidence that at both NASA's Johnson Space Center in Houston, Texas, and at the Arecibo Observatory on the island of Puerto Rico, dark, cloaked monsters were seen and went on to become associated with the violent killing of animals, and the subsequent draining of blood from their corpses, in the immediate vicinities? Perhaps not, as we shall now see.

For years, sensational and sinister stories have surfaced from the forests and lowlands of Puerto Rico that tell of a strange and lethal creature roaming the landscape by night and day, while striking overwhelming terror into the hearts of the populace—not at all surprising, as the animal has been described as having a pair of glowing red eyes, powerful, claw-like hands, razor-sharp teeth, a body not unlike that of a monkey, a row of vicious spikes running down the length of its back, and occasionally large and leathery bat-like wings. And if that is not enough, the beast is said to feed on the blood of the local animal (predominantly goat) population. Puerto Rico has a monstrous vampire in its midst. Its name is the chupacabra, a Latin term, very appropriately meaning "goat-sucker."

Theories abound concerning the nature of the beast, with some researchers and witnesses suggesting that the monster is some form of giant bat; others prefer the theory that it has extraterrestrial origins, and the most bizarre idea postulated is that the chupacabra is the creation of a top-secret genetics research laboratory hidden deep within Puerto Rico's El Yunque rainforest, which is located in the Sierra de Luquillo, approximately 25 miles southeast of the city of San Juan. And, yet again, we see NASA playing a part in the quest to understand the nature of this monstrous, winged, blood-loving animal killer.

On several occasions, I have traveled to Puerto Rico to try and seek out the vampire-like chupacabra for myself. And, on each and every occasion, I have heard intriguing rumors to the effect that in the mid-1980s a number of extremely violent chupacabra were captured by U.S. military personnel after the crash of a UFO on the island, and were transferred to a secret NASA facility in Florida. Is such a scenario just too fantastic for words? Certainly, not everyone seems to think so. From three serving and one retired Puerto Rican civil defense personnel I heard basically the same story: On the night of February 19, 1984, a spacecraft from a faroff world plunged deep into the heart of the El Yunque rain forest after suffering some form of midair calamity. The impressively sized craft was said to be circular in shape, white in color, and badly damaged by the force of the impact.

According to the story, the first to arrive on the scene was a unit of U.S. military personnel that was in the process of securing the site when, to its horror, it was confronted by five deadly chupacabra that quickly attacked and slaughtered seven of the military personnel as they valiantly sought to try and contain the situation. Reportedly, the creatures then quickly retreated into the bowels of their wrecked spacecraft, where they remained hunkered down until a unit of NASA scientists and a team of security personnel arrived, armed with makeshift cages. The plan was to try and recover what was left of the UFO and, if possible, take all of the beasts alive. Things did not go well at all: At least five of the security team and two of the scientists were killed, as were three of the bloodsucking creatures. The remaining two chupacabra, however, were successfully sedated, quickly caged, and flown back to the United States and the secure confines of an underground laboratory that fell under the auspices of NASA.

Should we dismiss the story as being merely a piece of folklore and mythology? Or, incredibly, could the events of February 19, 1984, have been all too real? It seems that, unless NASA elects to one day come clean, the answer to that question will remain frustratingly elusive. Of one thing, however, we can be pretty sure: Strange winged monsters that seem to act in vampire fashion and have a particular penchant for slaughtering animals seem to be curiously well-known to NASA—on the island of Puerto Rico and within the perimeters of the Johnson Space Center in Texas.

CLASSIFIED

I have saved until the very last what is quite possibly the absolute strangest, and certainly the most controversial of all the stories that posit a link between NASA and unusual, animalistic life-forms. The story came from a man named Bruce Weaver, whose grandfather worked at Brooks Air Force Base, Texas, in a medical capacity in the early to mid-1960s. On November 21, 1963—one day before President John F. Kennedy was assassinated in Dealey Plaza (Dallas, Texas) and nine days after JFK had instructed NASA Director James Webb to try and develop a program with the former Soviet Union in joint space and lunar exploration—a highly significant event occurred at Brooks AFB. On that day, President Kennedy dedicated six new aerospace medical research buildings there that were said to be vital to NASA's burgeoning manned space program. The dedication proved to be historic, in the sense that it was Kennedy's last official action in his elected role as president of the United States of America.

Interestingly, there have been rumors suggesting that JFK may possibly have seen far more than just the latest results in aerospace medicine during his visit to Brooks, such as the biological remains of alien cadavers recovered from the UFO crash at Roswell, New Mexico, in 1947. This story is made all the more intriguing by the fact that among those with whom JFK was arranged to meet at Brooks was Major General Theodore C. Bedwell, Jr., who, from 1946 to 1947, served in the position of Deputy Surgeon and Chief, Industrial Medicine, Air Materiel Command ,at Wright Field, Ohio—the precise location to where, it is alleged by many UFO investigators, the bodies from Roswell were taken midway through 1947, after their recovery from the harsh deserts of New Mexico.

Also during the visit to Brooks an arrangement was made for Kennedy to meet with a certain Colonel Harold V. Ellingson, of the U.S. Air Force, who had received a Bachelor of Science degree in Bacteriology in 1935. Ellingson held a number of posts, the most interesting being that of Post Surgeon and Hospital Commander at Fort Detrick, where, as was noted earlier in this book, research into alien viruses was reportedly undertaken, in part by NASA employees, during the 1970s. Another part of the itinerary planned for JFK on his visit to Brooks AFB involved the president receiving a briefing from the staff of the 6570th Aerospace Medical Research Laboratories, whose work centered upon the feasibility of NASA safely putting a man into outer space, or near-Earth orbit, from a medical and biological perspective.

Bruce Weaver's grandfather, Fred, was good friends with Major General Theodore C. Bedwell, Jr., and shared a passion with the general for aerospace medicine, the future world of manned space flight, and how humankind might best cope with the rigors of exposure to the somewhat poorly understood realm of outer space. But that is not all: Bruce Weaver states that a close colleague of Bedwell once confided in his grandfather that, during the course of the same program that culminated in the dedication by President Kennedy of six new aerospace medical research buildings at Brooks AFB in 1963, he had once read a classified report on, and had viewed a number of color photographs pertaining to the recovery of a highly unusual humanoid body by NASA security personnel at what is today known as the John H. Glenn Research Center at Lewis Field, Ohio (which is where the liquid hydrogen rocket engines were successfully developed that helped ensure NASA's Apollo astronauts reached the Moon). The body, said Weaver, had been secretly transferred to Brooks Air Force Base for study by Air Force and NASA personnel whose expertise was in the field of aerospace medicine.

From what Fred Weaver was told, two nights before the body was retrieved from woodland surrounding the Glenn facility, there had been a wave of UFO activity over the immediate area—most of which manifested in the form of small, fast-moving balls of blue light that zipped around the NASA installation and in the heart of the woods, and which gave all the indications of being either intelligently controlled, or perhaps even sentient, ethereal entities in their own right.

But it was the nature of the body itself that Fred Weaver found so amazing and unforgettable. According to what he was told by the colleague of Major General Bedwell, Jr., the recovered creature had been shot and killed by NASA security operatives who were engaged in keeping a close watch on the strange balls of floating light in the area. It was an immense, approximately 9-foot-tall, powerfully muscled monster. It was covered in short, coarse brown hair, and seemed to have attributes and features that were part ape and part human. In other words, NASA had gotten hold of nothing less than a corpse of what, today, pretty much everyone would associate with the legend of Bigfoot.

Weaver was informed that the body of the beast was autopsied and studied extensively. The most significant findings were:

- The Bigfoot had 32 teeth, as do most human beings.
- Its vocal chords seemed to be very human-like in appearance.
- Strangest of all, within its lower, left arm was embedded some form of small, metallic device that NASA medical specialists concluded was very likely a type of highly advanced tracking device and/or transmitter.

Who had placed the device inside the creature, and for what particular purpose? This was a question that could not be answered. Eventually, says Weaver, he heard that the remains of the hairy giant, including the curious device, were transferred to the Foreign Technology Division (FTD) at Wright-Patterson Air Force Base, Ohio, for additional study. The results of that study, unfortunately, remain unknown.

Bruce Weaver readily admitted that the story was both controversial and outrageous, but stood solidly by the words of his grandfather:

> Grandpa said it, and that's good enough for me. If it's not good enough for anyone else, that's too bad for them.[8]

Chapter 14
Opening NASA's X-Files

The Freedom of Information Act has proven to be a very useful tool when it comes to trying to ascertain what NASA may know about significant UFO events and alien visitations. Although no truly smoking gun has yet surfaced, a number of official reports that NASA has on file, which relate directly to the UFO controversy, tell startling and illuminating stories—particularly those that occurred in the mid- to late 1980s and mid-1990s.

On May 19, 1986, NASA documentation reveals, Brazil got the full brunt of what sounds like a near invasion force of UFOs. According to data provided to NASA by Department of State contacts in Brazil, on the night in question, a veritable fleet of UFOs—at least 20 flying at tremendous speeds—were independently seen by military aircrews of the Brazilian Air Force, ground radar operators of the military, and those of civilian airports.

According to NASA's files on the matter, the UFOs were first seen around 8:45 p.m. by the pilot of a Xingu aircraft that was transporting none other than Ozires Silva, the former president of Embraer, between San Paulo and Rio de Janeiro. As a result of the pilot's encounter, three fighter aircraft from Santa Cruz quickly took to the skies. At around 9 p.m., recorded NASA, the crews of each aircraft recorded anomalous targets on their radar screens, while one of the air crews described seeing red, white, and green lights performing astonishing movements in the sky.

Thing heated up significantly when radar contact was made with even more UFOs near Brazil. As a result, three Mirage aircraft were ordered airborne from Brazil's Anápolis Air Base and subsequently gave chase. At around 20,000 feet, NASA was told, all the aircrews made sudden and dramatic contact with the UFOs. No less than 13 disc-shaped objects, surrounded by red, green, and white lights, loomed into view and proceeded to "escort" the Brazilian Air Force pilots "at a distance of one to three miles," after which the UFOs shot away at high speed.[1]

Notably, the author of the report told NASA that:

> The Air Minister is quoted by the press as saying there were three groups of targets on the ground radar and that the scopes of the airborne radars were saturated." In addition, said the writer: "Three visual sightings and positive radar contact from three different types of radar systems, leads one to believe that something arrived over Brazil on the night of 19 May.[2]

With that specific, latter point, I can only heartily concur.

TOP SECRET

A document dated December 3, 1989, forwarded to NASA by the CIA, tells a truly remarkable story of an event that had occurred in the former Soviet Union less than one day after the destruction of the *Challenger* space shuttle in January 1986. At the time, it attracted the close attention of the Russian media. To what extent NASA's interest in the affair may have been prompted by the close proximity, time-wise, to the space shuttle catastrophe is unknown. But, that this might have had some bearing on NASA's decision to look into the Soviet case is at least worth keeping in mind.

The author of the document recorded:

> Setting the scene for the media coverage was an article in the 9 July 1989 Sotsialisticheskaya Industriya, which referred to many recent reports of UFO sightings in the USSR. Interviewed by the paper, P. Prokopenko, director of a laboratory for the study of "anomalous phenomena," stated that a "permanent center" for the study of UFOs is being established in the Soviet Union. In addition to conducting research and presenting lectures in UFOs, the center will support the investigation of reported sightings.[3]

The author of the report then elaborated upon the remarkable tale of a UFO crash incident that had reportedly occurred in the latter days of January 1986. According to the story, a sensational encounter had occurred at a location known locally as Hill 611, which was (and still is) situated close to the village of Dalnegorsk in Primorskiy Kray.

According to what NASA was told, although the incident was still the subject of deep and intensive study, investigators had been able to discern that numerous people living in the vicinity of Hill 611 were witness to "a flying sphere crash into one of the hill's twin peaks." Moreover, a number of unnamed physicists and "other scientists from the Siberian Division of the USSR Academy of Sciences" were reportedly studying a curious "fine mesh," "small spherical objects," and "pieces of glass" that were found at the site of impact, and that were considered to be "small remnants left behind by the sphere," which had been all but obliterated as a result of the earth-shattering impact.[4]

The report continued:

> In studying the site, a scientist A. Makayev reported finding gold, silver, nickel, alpha-titanium, molybdenum, and compounds of beryllium. One of the "skeptical" physicists from Tomsk has hypothesized that the so-called sphere could have been some kind of a "plasmoid," formed by the "interaction of geophysical force fields," which captured the elements found by Makeyev from the atmosphere

> on its trajectory toward disintegration on the hilltop. Other researchers have generally rejected this explanation since the amounts of various types of metals found at the site would imply, according to this "plasmoid" theory, that "the concentration of metals in the atmosphere should exceed the present level by a factor of 4,000."[5]

Inevitably, there were those researchers and investigators who had succeeded in reaching the crash site and believed the mysterious sphere that plunged to Earth on Hill 611 was "an 'extraterrestrial' space vehicle constructed by highly intelligent beings." Indeed, the report to NASA referred to a Doctor of Chemical Sciences, one V. Vysotskiy, who stated on the record that "without doubt, this is evidence of a high technology, and it is not anything of a natural or terrestrial origin."[6]

Not everyone was in agreement with Vysotskiy, however, as the document noted: "Physicist Yuriy Platov of the Terrestrial Magnetism Institute does not believe the claims of scientists who maintain they have found remnants in Dalnegorsk of a UFO constructed by extraterrestrials." Rather, Platov was of the opinion that the debris and materials were the result of a catastrophically unsuccessful, secret Russian rocket launch somewhere in the area; while at least some of the unusual activity might have been due to a rare, natural weather based phenomenon known as ball-lightning.[7]

According to additional memoranda made available to NASA, another Russian scientist had also commented on the alleged crash of a UFO at Dalnegorsk—and on the reported UFO crash at Roswell, New Mexico, in the summer of 1947:

> The article contrasted Platov's view with that of another physical scientist, Vladimir Azhazha, who was recently elected chairman of the All-Union Commission for the Study of Unidentified Flying Objects of the Union of Scientific and Engineering Societies. Azhazha compared reports of a UFO crash in the USSR with a claim by UFO enthusiasts in the United States that a UFO had crashed in the desert

near Roswell, New Mexico, in 1947. He believes there is sufficient evidence to support the claims of UFO crashes in both cases—in Dalnegorsk and in Roswell. In the latter case, he cited the testimony of eyewitnesses who maintained that they had seen the bodies of four extraterrestrials lying near the smashed spacecraft. According to Platov, however, the eyewitnesses in the Roswell case were mistaken. He believes that the object that crashed was a USAF experimental rocket with four Rhesus monkeys aboard and that the accident was the result of an unsuccessful launch attempt at the dawn of the space era.[8]

Was Roswell really the result of an early secret experiment at the beginnings of the space age, rather than something that involved aliens? Is this, perhaps, why NASA, as the agency at the forefront of research into the arena of outer space, was kept informed of Azhazha's conclusions? We may never know; an attempt to secure further information from NASA on this particular matter has been met with nothing but a denial of any further pertinent documentation.

CLASSIFIED

One of the strangest, and certainly most unsettling reports that found its way into the inner sanctum of NASA in the decade of the 1990s concerned an unusual event that occurred in Somaliland in January 1996. A three-page document, shared with NASA by the CIA, titled "Somaliland President Egal Speaks on Mysterious Bomb Blast," details a series of strange and unidentified explosions that had the recently occurred in the region, which some local commentators were directly attributing to the presence of UFOs.

The available documentation that NASA has now made available also reveals that the physical and mental health of both animals and people in the region had been adversely affected by whatever it was that had exploded in Somaliland. Symptoms included dementia, bodily rashes, boils on the skin, aching stomachs, and even skin-shedding. According to files in the possession of NASA, the unidentified aerial object that had either crashed or exploded in the area had been moving at supersonic speed. As a result of the sheer size and remoteness of the area, a search to locate and recover any potential debris from the device had proven unsuccessful.

Perhaps most significant of all, however, is the fact that documentation made available to NASA recorded the intriguing words of Somaliland's President Egal, who proved to be very vocal on the matter of the high strangeness that was afoot in his country:

> We have these mysterious reports from our nomadic population there. And then, I sent a four-man commission, and they have submitted to us a report, which is very, very alarming. Most of the animals in the area are still in a sort of a demented stage.[9]

Today, the curious events of January 1996 remain as mysterious as they were at the time. That is, unless NASA knows better.

Chapter 15
Roswell and the Astronaut

On July 8, 1947, the New Mexico–based *Roswell Daily Record* newspaper announced in bold, unbelievable headlines that staff at the nearby Roswell Army Air Field (RAAF) had recovered the remains of nothing less than a crashed, unidentified, flying object. Under the headline of "RAAF Captures Flying Saucer on Ranch in Roswell Region," the newspaper ran with the words of an official press release that the RAAF's Press Information Officer, a man named Walter Haut, had issued only hours earlier:

> The many rumors regarding the flying disc became a reality yesterday when the Intelligence office of the 509th Bomb Group of the Eighth Air Force, Roswell Army Air Field, was fortunate to gain possession of a disc through the cooperation of one of the local ranchers and the sheriff's office of Chaves County. The flying object landed on a ranch near Roswell sometime last week. Not having phone facilities, the rancher stored the disc until such time as he was able to contact the sheriff's office, who in turn notified Maj. Jesse A. Marcel of the 509th Bomb Group Intelligence Office. Action was immediately

taken and the disc was picked up at the rancher's home. It was inspected at the Roswell Army Air Field and subsequently loaned by Major Marcel to higher headquarters.[1]

The undeniably mysterious event at Roswell has subsequently become the subject of dozens of nonfiction books, a number of science-fiction novels, official studies undertaken by both the General Accounting Office and the U.S. Air Force, a plethora of television documentaries, a movie starring Martin Sheen and Kyle Mclachlan, and considerable media scrutiny of both a pro and con nature. The affair has also left in its wake a near mountain of theories to explain the event, including:

- ✓ A weather balloon
- ✓ A Mogul Balloon secretly utilized to monitor for Soviet atomic-bomb tests
- ✓ An extraterrestrial spacecraft
- ✓ A series of dark and dubious high-altitude-exposure experiments using Japanese prisoners of war
- ✓ Some sort of near-catastrophic atomic mishap
- ✓ The crash of a Nazi V-2 rocket with shaved monkeys onboard
- ✓ An accident involving an early "Flying-Wing" style aircraft, secretly built by transplanted German scientists who had relocated to the United States following the end of the Second World War.

Whatever the truth of the matter, and the ultimate point of origin of the craft and its strange crew, it is an undeniable fact that the military hastily, and decisively, retracted its sensational statement that had appeared on the front page of the *Roswell Daily Record*, preferring, instead, to substitute it for a far more down-to-earth and prosaic one, in which it was asserted that the materials found near Roswell, and then subsequently brought to the Roswell base, originated with nothing stranger than an ordinary, mundane weather balloon.

In an effort to try and add further weight to this new, albeit far less exciting, scenario, photographs showing Brigadier General Roger Ramey, of the Eighth Air Force at Fort Worth, Texas, surrounded by what were quite obviously everyday balloon-based materials, were taken, and paraded for one and all to see. The press of the day, all around the globe no less, unfortunately totally bought into the cover-up—hook, line, and sinker. Stranger still: no one within the media seemingly ever thought to ask the military an obvious and glaring question: Why were highly trained military personnel of the Roswell Army Air Field initially unable to differentiate between something as fantastic and futuristic as a flying saucer and something as normal as a weather balloon?

TOP SECRET

For approximately three decades, that question remained wholly unanswered—that is, until UFO researcher and nuclear physicist Stanton T. Friedman began to dig deep into the heart of the mystery. It was January 20, 1978, and Friedman was lecturing in Baton Rouge, Louisiana. While there, he took part in a variety of interviews with local media. It was during an interval in one such interview, at a television station in town, that Friedman was introduced to the station's manager, who just happened to be a good friend of the still surviving Jesse Marcel, and who, according to the official press release from the RAAF of July 8, 1947, was a key player in the recovery of the strange materials found out in the New Mexican desert on that famous day. Thus began Friedman's intense, driven quest—a quest that still continues to this day—to learn the truth about what really occurred at Roswell.

Welcome to Roswell, New Mexico: The home of crashed UFOs.

When interviewed, Marcel was willing to admit that he had viewed at the site a great deal of unidentified debris, which covered an estimated area of three quarters of a mile by several hundred feet, but stressed that debris was *all* he saw. In other words, there were no bodies, and no intact, or even semi-intact vehicle (at least, none that he ever saw). It was Marcel's opinion that the object, whatever it may have been, had almost certainly exploded high in the sky, and the wreckage had then rained down across the desert floor.

As for what the debris actually seemed to represent, Marcel said:

> I was pretty well acquainted with most everything that was in the air at that time, both ours and foreign. I was also acquainted with virtually every type of weather-observation or radar-tracking device being used by either the civilians or the military. What it was we didn't know. We just picked up the fragments.[2]

Marcel was certain, however, that the debris was unlike anything he had ever seen before or since, and noted that "it certainly wasn't anything built by us."[3]

CLASSIFIED

As Friedman's research progressed onward and upward, he came into contact with another investigator, William L. Moore, who had uncovered intriguing strands of what sounded like the same event, but from the perspective of additional witnesses, of whom Friedman was then unaware. Friedman and Moore duly began to share their data, and were able to determine that the location where the crash and subsequent recovery had occurred was a very isolated piece of farmland known as the Foster Ranch. Situated about 75 miles north of the town of Roswell, at the time the ranch was worked by William Ware "Mac" Brazel, who, having died in the 1960s, was unfortunately not available for interview. But, very fortunately, there were others who remembered that heady period very well indeed.

Brazel's son, Bill, for example, related a highly intriguing account concerning his personal recollections of the nature of the weird materials recovered at the ranch. He conceded that it seemed to resemble regular tinfoil in some ways, except for one significant way: It couldn't be torn, at all. He added:

> You could wrinkle it and lay it back down and it immediately resumed its original shape. [It was] almost like a plastic, but definitely metallic in nature. Dad once said that the Army had once told him it was not anything made by us.[4]

As time progressed, retired Major Jesse Marcel, perhaps finally feeling more relaxed about speaking publicly on the affair after others started doing likewise, began to divulge further data on the strange debris found out in the blistering heat of the New Mexico desert:

> [It] could not be bent or broken or even dented by a 16-pound sledgehammer. [It was] almost weightless, like a metal with plastic properties.[5]

Yet another jaw-dropping story of great significance to the controversy of what did or did not crash in New Mexico in July 1947 came from Vern and Jean Maltais, a married couple who stated that an old friend of theirs, a man named Grady Barnett, a field engineer attached to the Soil Conservation Service, had guardedly informed them that at some point in the summer of 1947 he had stumbled upon the remains of a very unusual-looking vehicle on New Mexico's Plains of San Augustine that was disc shaped and appeared to be constructed out of something that resembled stainless steel. Sprawled around the object were a number of dead bodies of a very unusual nature, Barnett told Vern and Jean Maltais: All were short, with oversized, bald heads, and strangely spaced eyes.

Barnett stood staring for a while at the shocking scene, shocked and unsure of what he should do. That is, until a detachment of military personnel quickly arrived on the scene and sternly told him exactly what to do: Say nothing to anyone about what you have seen, or else. Barnett, for a while at least, got the military's message loud and clear.

In 1980, the Roswell case was taken to a whole new level when William Moore coauthored with Charles Berlitz a full-length book on the case that was titled, quite appropriately, *The Roswell Incident*. At the time that the Berlitz-Moore book was published, most of the available evidence in hand was that which had been uncovered by Friedman and Moore. And, five years later, the ongoing research of the industrious pair had led to the identification of, astonishingly, almost a hundred people who were implicated in the events of July 1947, to varying degrees of significance and relevance.

A new development of some particular note occurred in July 1985, when Moore spoke at the annual symposium of the Mutual UFO Network (MUFON) in St. Louis, Missouri. Moore revealed the intriguing story of what had taken place when Friedman had carefully broached the alien-body angle of the Roswell story with a man named Lewis Rickett, who had been stationed at Roswell in 1947 with the ultra-secret Counter Intelligence Corps (CIC). In Friedman's words, when the matter of the bodies was brought up in the conversation:

> Rickett clearly reacted and indicated that this was an area he couldn't talk about. He indicated there were different levels of security about this work; that a directive had come down placing this at a high level. He went on to say that certain subjects were discussed only in rooms that couldn't be bugged.[6]

CONFIDENTIAL

The industrious team of Moore and Friedman was not the only one looking into the complexities of the Roswell story. Additional data, of an equally startling nature, began to surface in the late 1980s, and continued well into the 1990s, from researchers Kevin Randle and Donald Schmitt. The collective data the pair succeeded in uncovering told what was undoubtedly a highly provocative story: It was reportedly on the night of July 4, 1947, that a stricken vehicle from another world crashed outside of Roswell, having already disgorged a considerable amount of material on the desert floor of the Foster Ranch, following, one presumes, some form of dire, midair calamity. The object was not of a classic saucer shape, however. Rather, Randle and Schmitt related that it was somewhat narrow with what was described as a bat-like wing, and was around 30 feet in length—these points were particularly stressed by the otherwise relatively tight-lipped Counter Intelligence Corps operative, Lewis Rickett.

Interestingly, also according to Rickett, in September 1947 he became embroiled in the Roswell affair at a deep and notable level when he spent a period of time working with Dr. Lincoln La Paz of the University of New Mexico. The story that Rickett told as it related to Roswell was that he and La Paz were assigned to specifically determine the speed and trajectory of the vehicle that had crashed. Interestingly, the pair found what was interpreted to be a possible touchdown point, where, perhaps, the vehicle had made an emergency landing before complete disaster finally struck. This site was described as being around five miles from where rancher Brazel located the strange materials on the Foster Ranch, and where Major Jesse Marcel, Sr., had been present. Amazingly, at the site of the touchdown, the sand had reportedly been crystallized as the result of what was deemed to be exposure to an extraordinarily high temperature. Significantly, Rickett recalled that La Paz had prepared an official report on their collective findings for the attention of cleared personnel in the Pentagon that offered the theory that what had crashed was very possibly some form of unmanned probe from another world.

That La Paz was apparently not briefed in advance on the fact that unusual-looking bodies were also said to have been recovered (hence his conclusion that the device was unmanned) suggests strongly that, as Rickett had very guardedly advised Friedman was the case, there were different levels of security surrounding the Roswell event, and that "need to know"—or *not* to know—was an overwhelming factor in the extent to which official players in the story were briefed on certain integral aspects of what had reportedly occurred.

Randle and Schmitt concluded, based upon the testimony of their sources, that as many as five alien bodies may have been found in the desert, one of which, for a while, might have survived the crash. Here are some of their accounts:

- ✓ Edwin Easley, the Provost Marshal at Roswell, made vague death-bed comments about the creatures.
- ✓ Sergeant Melvin E. Brown stated that he saw bodies that looked Asian in appearance, had large heads, no hair, and a yellowy tinge to their skin.
- ✓ Chillingly, a death threat was made to the family of Frankie Rowe, whose father served with the Roswell Fire Department at the time of the crash, and who was also implicated in the

controversy to some degree. According to Rowe, her father quietly confided in her family that he was a witness to unusual materials and strange bodies, as well as to one, solitary being that had survived the desert impact: "The one that was walking was about the size of a 10-year-old child, and it didn't have any hair…it seemed so scared and lost and afraid." Rowe adds that after the events in which her father was implicated had occurred, a number of military personnel came to the family home and made it abundantly clear that if anyone said even a word about the Roswell events, "they might just take us out to the middle of the desert and shoot all of us and nobody would ever find us."[7]

- Reporter Johnny McBoyle's account also fits the pattern of events. McBoyle claimed to have seen an impressively sized, dented device at the crash site that resembled a dish pan. He reported back to Roswell's KSWS Radio Station that rumors were floating around that the military was going to recover the device—along with a number of bodies of what McBoyle described as little men. Notably, McBoyle later declined to speak further about what it was that had really occurred in the New Mexican desert.

CLASSIFIED

Despite the wealth of testimony and data that surfaced between 1947 and the late 1990s, the U.S. government steadfastly and stubbornly ignored all of the claims that had been made about Roswell in a specifically UFO context. That is, until 1994, when, in response to persistent and dogged questioning from the late congressman for New Mexico, Steven Schiff, the United States Air Force grudgingly conceded that a mundane weather balloon was not the real culprit behind the Roswell controversy, after all. Rather, asserted the Air Force, the device that had come down on the Foster Ranch was a balloon that originated with a classified operation known at the time as Mogul. It was a secret operation that utilized balloons to carry radar reflectors and acoustic sensors aloft for the purpose of determining the current state of Soviet atomic weapons research.

Then, in 1997, in what was certainly a surprising and wholly unanticipated move, the Air Force finally addressed, in a full-length report that ran to a couple of hundred pages, no less, the controversy surrounding the allegations that unusual, or potentially even alien bodies had been found in the New Mexico desert. The Air Force's conclusions were that people had mistakenly seen crash-test dummies utilized in parachute-based experiments. The fact that such experiments did not even begin until the early 1950s did not faze the Air Force in the slightest, which arrogantly and outrageously proclaimed that all the firsthand witnesses to the supposed bodies had consistently mistaken the year in question due to what officials ingeniously termed "time compression."

Essentially, that is where matters stand today: The collective UFO research community largely, although not unanimously, stands by its views and conclusions that aliens met their deaths in the harsh deserts of New Mexico more than 60 years ago, while the Air Force asserts that nothing stranger than a Mogul balloon and crash-test dummies were the real culprits. However, that same UFO research community has a powerful and famous ally that the Air Force utterly lacks—one of NASA's Apollo astronauts: Dr. Edgar Mitchell.

CONFIDENTIAL

Born in 1930, Edgar Mitchell piloted the *Apollo 14* lunar module that touched down on the Moon on February 5, 1971. As well as being one of the very few people to walk on the lunar surface, Mitchell is a firm believer that UFOs are real, and that the truly unknown ones are the handiwork of intelligent entities from other worlds—aliens, in other words.[8] Moreover, Mitchell has said that UFOs have been the subject of officially

NASA astronaut Edgar Mitchell, shown here on the moon, believes that aliens crashed at Roswell in 1947.

orchestrated and disseminated disinformation in order to try and deflect attention away from the UFO controversy, and to sow the seeds of confusion as part of an attempt to prevent the true picture from ever publicly surfacing. *Dateline NBC* interviewed Mitchell on April 19, 1996, during the course of which the astronaut revealed that he had personally met with officials from three countries who claimed to have had personal encounters with extraterrestrials. He went on to express his view that the evidence for alien contact was without doubt overwhelming.

In an exclusive interview undertaken on October 10, 1998 from his Florida home, Mitchell told journalist John Earls, of Britain's *People* newspaper, that he was certain that extraterrestrial life is a reality, and that he believed at least some of that same alien life has visited the Earth. Mitchell added the following to Earls:

> As a former astronaut, the military people who have access to these files are more willing to talk to me than to people they regard as mere cranks. The stories I have heard from these people...leave me in no doubt that aliens have already visited Earth.[9]

In 2004, Mitchell went a step further and publicly stated that an elite group that existed within the murky world of U.S. officialdom was secretly investigating UFO phenomena, and had in its possession the bodies of an unspecified number of dead extraterrestrials that, presumably, had been recovered from one or more UFO crashes. He said to the *St. Petersburg Times*:

> We all know that UFOs are real; now the question is where they come from.[10]

It was four years later, however, when Mitchell really made the news with respect to his views on UFOs and alien life. On July 23, 2008, while being interviewed by Nick Margerrison on Britain's *Kerrang Radio*, Mitchell said that aliens had indeed crashed at Roswell, New Mexico, in the summer of 1947, and added that there had been a concerted effort by the authorities to suppress the facts.

> I happen to have been privileged enough to be in on the fact that we've been visited on this planet, and the UFO phenomenon is real.[11]

It did not take NASA long to formulate a reply to these fantastic statements from one of its own Apollo astronauts. NASA clearly found itself in a quandary: There was certainly no desire on NASA's part to cast any doubt on Mitchell's status as an American hero, but the space agency realized that it had to distance itself from such claims of the unearthly kind. Very diplomatically, a spokesperson merely said that Mitchell's opinions were not NASA's opinions:

> NASA does not track UFOs. NASA is not involved in any sort of cover-up about alien life on this planet or anywhere in the universe. Dr Mitchell is a great American, but we do not share his opinions on this issue.[12]

Two days after the *Kerrang* interview aired, Mitchell was interviewed by *Fox News*, where he was careful to clarify to the host that his comments regarding his personal knowledge of UFOs did not extend to NASA. He was, however, willing to quote unnamed sources, now dead, at Roswell, who had personally confided in him that the controversy of July 1947 did indeed have at its heart the crash and recovery of a spacecraft from another world. Mitchell also stated that he had received confirmation of the reality of the Roswell saga from a presently unnamed intelligence officer based somewhere in the heart of the Pentagon.

And still the interviews kept on coming: The floodgates, not surprisingly, were now well and truly open, and were showing no signs of closing in the immediate future. In the wake of the sensational *Kerrang* interview, Mitchell told Lisa Bonnice, host of the *Shape-Shifting* radio show, that, because he had grown up in the Roswell area and had later traveled to the moon on *Apollo 14*, some of the old-timers from that period, as well as military and intelligence people who were under "rather severe oaths" to not reveal anything relative to Roswell "wanted to get their conscience clear and off their chests before they passed on."[13]

Still on the subject of Roswell, Mitchell explained to the Discovery Channel that, with respect to what he had been told about the crash at Roswell, he did take his story to the Pentagon "and asked for a meeting with the Intelligence Committee of the Joint Chiefs of Staff, and got it. I told them my story, and what I know, and eventually had that confirmed by the admiral that I spoke with, that indeed what I was saying was true."[14]

Despite the fact that Mitchell is a true American hero, one who risked his life to travel to the moon, and who has significantly added to the United States's place in world history in the process, it is truly unfortunate that his views on UFOs and the Roswell affair are summarily dismissed by both NASA and his very own government. It is a near certainty that, had he been speaking about any other aspect of his life and his time as a NASA astronaut, Dr. Edgar Mitchell would have received far better and respectful treatment than the swift dismissal that NASA offered in response to the public airing of his views on the Roswell, New Mexico, UFO crash of July 1947. Such, it seems, is the profound and ongoing effect of the huge blanket of secrecy that still surrounds this decades-old case.

Chapter 16
Censored Photos

In March 1960, NASA became deeply embroiled in a controversial UFO event involving a man named Joseph Perry of Grand Blanc, Michigan, who claimed to have photographed a UFO in the night sky approximately two weeks earlier. Perry was interviewed by agents of the Detroit office of the FBI on March 5, 1960, who in turn forwarded their files on the strange affair to the Air Force Office of Special Investigations at Selfridge Air Force Base, and also to senior personnel at NASA headquarters. Interestingly, a handwritten note from a NASA source to the FBI on the matter states: "Please do what you can to obtain Mr. Perry's photograph for our interest. An opportunity to examine would be appreciated."[1]

Perry, who at the time operated a pizza restaurant in Grand Blanc, had "been a professional photographer for 30 years," and would often take pictures of the moon with his homemade telescope, as the FBI reported to NASA. At around 1 a.m. on February 21, 1960, Perry told the FBI—and, by default, NASA too—that he took a number of photographs of the moon, duly developed them in his own darkroom, and while looking at them later, was astonished to notice what appeared to be "a flying object somewhere between the end of his telescope and the surface of the Moon." A highly excited Perry quickly had the picture blown up, which made the object much clearer. As Perry carefully scrutinized the new version of the picture, he could see that it appeared to show a clearly delineated, structured vehicle that was oval-shaped, had a flat bottom, seemed to be surrounded by a fluorescent glow, and even appeared to have "a vapor trail running behind it." The FBI advised NASA that Perry "has taken over 1,000 pictures of the Moon and has never seen anything resembling this object."[2]

NASA told the FBI to please do anything it could to secure either the original or a copy of Perry's photograph of the UFO. Certain that he had captured something truly anomalous on film, Perry duly furnished the FBI with the relevant, original photograph, which then sent it to the aforementioned Office of Special Investigations at Selfridge Air Force Base. From there, it made its way to the heart of NASA—and vanished.

By this time, the media had latched on to the story: the Michigan-based *Flint Journal* splashed details of the affair, in bold headlines, across its pages. Notably, the newspaper chose to quote the National Investigations Committee on Aerial Phenomena (NICAP), which had informed Perry, much to his concern, that:

> From past experience with photographic evidence we consider it unlikely that you will ever see your picture again.[3]

Perry responded by contacting the FBI, which merely commented that the picture was now in the "proper hands," to which Perry responded:

> The only way I will be satisfied if I don't get it back is if the government tells me it is top secret.[4]

In the weeks and months that followed, a considerable file was built up by NASA and the FBI about Perry's photograph, much of which dealt with his concerns about trying to retrieve his property from officialdom.

Matters were resolved (to the satisfaction of NASA, the FBI, and the Air Force Office of Special Investigations at least) when, according to the official documentation on the affair, Perry was duly advised that "what appeared to be a flying object in this slide is actually a part of the negative which was not properly developed."[5] Naturally, this statement did little to satisfy those who insisted that the photograph genuinely showed some form of structured vehicle that had its origins on another world—particularly so Joseph Perry, who viewed the whole situation through mystified and suspicious eyes.

Possibly recognizing that this controversy was going to run and run unless quick steps were taken to curb the situation, NASA and the FBI both insisted to all inquirers that this was a matter for the Air Force, and only the Air Force, and thus duly steered all incoming letters in the direction of the Air Force. In time, the controversy surrounding the matter of Joseph Perry's missing photograph, and the deep interest exhibited in it by NASA, faded away. Perhaps, one day, NASA will reveal more on this curious matter, and on the issue of why it was so interested in getting its hands on Perry's enigmatic picture. Or, maybe, we won't get that lucky. But, this was not the only occasion upon which NASA exhibited an interest in UFO photographs that then seemingly vanished into the ether.

TOP SECRET

Steven Greer is a well-known figure within the arena of UFO research, who has penned a number of books and organized large-scale events devoted to highlighting and exposing the issue of government secrecy in relation to UFOs. In 1990 Greer founded the Center for the Study of Extraterrestrial Intelligence, and is the brainchild behind what has become known as the Disclosure Project, described in the following fashion at the organization's Website, *www.disclosureproject.org*:

> The Disclosure Project is a nonprofit research project working to fully disclose the facts about UFOs, extraterrestrial intelligence, and classified advanced energy and propulsions systems.

Greer adds that the group has now secured the on-the-record testimony of more than 400 people—including numerous retired and former government employees, military personnel, and intelligence operatives—who have been brave enough to divulge the details of their personal firsthand knowledge or experience of profound UFO incidents, the attempts of the government to keep the subject hidden from the public, and even human interaction with aliens.

At the core of Greer's project are a number of central beliefs and acceptances:

1. Aliens exist.
2. They have been, and still are, clandestinely visiting the Earth.
3. Certain elements of the official world have acquired highly advanced extraterrestrial technologies and energies (possibly resulting from the crash, recovery, and secret study of one or more alien spacecraft).
4. The existence of these same technologies and energies could revolutionize our world, and, if placed into the public domain, could put an end to many (if not all) of the problems posed by our overwhelming reliance upon oil.

A desire to not adversely affect the status quo, the global economy, and the powerful corporations and individuals that wield power in the oil-producing industry are all seen by Greer and those who support and champion the work of the Disclosure Project as just some of the viable reasons why such extraordinary energies have not yet made their way into the heart of mainstream society.

Greer has been highly active, vocal, and forthright in trying to get his controversial message across to the world's press, to the general public at large, and even to the movers and shakers in the United States Congress. To that end, in May 2001, Greer held a historic press conference at the National Press Club, at which a number of people with official backgrounds—including some with past ties to NASA—spoke out about the UFO phenomenon and its links to officially orchestrated secrecy and conspiracy.

One of those who testified at Greer's 2001 event was a woman named Donna Hare, who had previously worked for Philco, a company with which NASA contracted at the dawning of the 1960s to provide NASA with a worldwide tracking station network for its manned Project Mercury flights. Hare, who became the recipient of a number of awards from NASA, including an Apollo Achievement Award, ultimately spent more than a decade and a half subcontracted to the space agency, working in Building 8 in the Photographic Laboratory at the Houston, Texas–based Johnson Space Center.

NASA's Lyndon B. Johnson Space Center, Texas, home to censored UFO photographs and a Mothman-type creature.

Hare had an enlightening story to tell the audience at Greer's 2001 event that strongly suggested certain senior elements within NASA knew that UFOs were an undeniable reality, but were doing their utmost to prevent such truths from ever becoming known outside of NASA's strict, regulated channels. Hare revealed to those present in the audience that, being in possession of a secret clearance at the time, she had pretty much open-door access, which, on one occasion, led to a memorable encounter. She had entered one particular room where much of the photographic data that had been collected on the Apollo missions to the moon, and by NASA satellites in Earth orbit, was dispatched to be developed and studied.

While there, Hare had the opportunity to speak with a colleague who showed her (perhaps, it might be argued, in violation of NASA's accepted, secret protocol) one photograph that displayed what was clearly a circular-shaped object in the skies. The man admitted to Hare that although he couldn't tell her what it did show, he could confirm that a particular part of his job involved him having to airbrush such imagery, specifically to ensure that aerial anomalies like this one never, ever made their way into the public domain. Someone (or some group) in NASA, it seems, had sent the order down for its own UFO photographs to be censored. And this would not be the only occasion when such a thing reportedly happened.

A further individual who made intriguing and, arguably, even more sensational statements relative to NASA's hidden UFO-related pictures was Karl Wolf, a U.S. Air Force operative whose clearance level was Top Secret. Midway through 1965, Wolf spent some time working on an assignment at Langley Air Force Base, Virginia, that was connected with NASA's lunar orbiter, and got talking to a fellow airman who was also working on the project. As the two chatted, it became clear that the airman was deeply worried about something big. He finally confided in Wolf that during the course of its analysis of its photographs of the moon, NASA had uncovered evidence of a large base of unknown origin and intent that was situated upon on the far side of the moon's surface. There was absolutely no doubt that those whose job it was to analyze the pictures were *not* misinterpreting unusual rock formations. Rather, the pictures clearly showed intelligently designed and crafted structures that made the mysterious base sound much more like a huge, sprawling space city than anything else.

Wolf, realizing that he had just been exposed to a story that was surely the subject of a high degree of official secrecy and stringent security, chose to break off the conversation, even though it had clearly fascinated him. For some time afterward, Wolf wondered if such a fantastic discovery would one day be announced to the world at large on the nation's news. It was not, of course. Just as in Las Vegas, what happens in NASA, it seems, is firmly destined to stay in NASA.

Clearly, we are seeing a notable pattern at work here: It is a pattern that was certainly in place at the time Joseph Perry took a photograph through his telescope of what some believed to be a UFO at Grand Blanc, Michigan, in 1960. And it seems to have still been in place years later, as the testimony of both Donna Hare and Karl Wolf collectively suggests. The censoring by certain elements of NASA of UFO-connected photographs has been going on for a very long time.

Chapter 17
Hacking NASA

The biggest stumbling block of all when it comes to verifying such intriguing stories is the complete inability on the part of the general public, the media, and the UFO research community to secure access to the priceless censored photographs for independent scientific scrutiny and evaluation. For its part, not surprisingly, NASA categorically denies the existence of any such photographs in the first place—of UFOs, or of some advanced, extraterrestrial base or structures on the surface of the moon. Similarly, nothing meaningful on this issue has ever surfaced into the public domain via the Freedom of Information Act.

In other words, therefore, the front door is firmly closed to anyone and everyone who does not possess the necessary official clearance to access the top-secret truth. That is, at least, unless you are prepared to go totally out on a limb, break the law to circumvent that front door, and instead penetrate NASA's back door via a series of brute-force computer hacks—which is precisely what has by now been done on a number of occasions. Indeed, a British man named Gary McKinnon did exactly this a few years ago, with truly alarming and disastrous results for the man.

A child of the 1960s, McKinnon currently lives under the threat of extradition to the United States on charges of carrying out what one U.S. prosecutor has claimed to be without doubt the biggest computer hack of the United States's official infrastructure in its history, causing severe damage to a whole host of NASA-, defense-, intelligence-, and military-operated computer systems. In his defense, McKinnon has repeatedly stated that he is only guilty of searching for top-secret UFO data—information of a precise type that would provide much weight for the revelations of Donna Hare concerning NASA's censored UFO pictures, and the beliefs of the Disclosure Project that the U.S. government has, for years, been deliberately suppressing extraordinary technologies that might very well have the ability to end our reliance upon oil, and radically alter the world (for the better) if made available to the populace at large.

But before turning our attention toward NASA and the McKinnon affair to an in depth degree, it is instructive to note that previous attempts have been made to hack into the systems of NASA and other elements of the U.S. government, military, and intelligence community, in search of secret UFO data and advanced, almost magical technologies. On each previous occasion, however, the perpetrators successfully avoided any form of punishment—a situation that American authorities were determined to avoid when it came to Gary McKinnon and his NASA intrusions.

TOP SECRET

Of the many rumors and allegations that surround the ongoing UFO controversy, one absolutely refuses to go away, despite the best attempts of the U.S. Air Force: At Wright-Patterson Air Force base in Dayton, Ohio, there exists a series of secret rooms, aircraft hangars, and underground chambers where the preserved remains of a number of dead alien creatures are stored, along with the wreckage of their crashed and recovered UFO. Further rumors borne out of such accounts imply that the fantastic alien technology that quite literally fell into the hands of the U.S. government has now been secretly duplicated, and the U.S. military is busy at work secretly building and flying its very own prototype UFOs. In generic terms, the location of this astonishing, fantastic evidence has become known as Hangar 18. Officially, the U.S. government and the Air Force vehemently deny that such sensational stories have any basis in fact, but an impressive body of testimony from a number of high-ranking (and on the record) sources strongly suggests otherwise.

The late U.S. senator Barry Goldwater, for example, said on the record in 1975 that:

> The subject of UFOs is one that has interested me for some long time. About 10 or 12 years ago I made an effort to find out what was in the building at Wright-Patterson Air Force Base where the information is stored that has been collected by the Air Force, and I was understandably denied this request. It is still classified above Top Secret.[1]

An equally thought-provoking account came from a former executive assistant to the deputy director and special assistant to the executive director of the Central Intelligence Agency: a man named Victor Marchetti. Revealing that during his time with the CIA, UFOs came under the category of very sensitive activities, Marchetti stated that he personally heard from high levels within the CIA that the bodies of little gray men, whose UFO had crashed, were secretly being stored at the Foreign Technology Division of Wright-Patterson Air Force Base. It is accounts such as these that led certain individuals in the computer-hacking world to seek out the answers for themselves via decidedly hazardous routes.

CLASSIFIED

On October 27, 1992, *Dateline NBC* devoted a segment of its show to the subject of computer hacking, and chose to include in that same segment select interviews with a number of self-confessed hackers. With one of the hackers talking about his ability to easily break into both government and military computer systems, NBC flashed across the screen a variety of documentation that had apparently been obtained by the hacker from Wright-Patterson's computer system, which in part stated, "WRIGHT-PATTERSON AFB/Catalogued UFO parts list, an underground facility of Foreign...." At that point, frustratingly, the camera panned away and the remaining segment of the material was not referenced in text format; however, it was later revealed that at least part of the material downloaded by the hacker was said to reference top-secret alien-autopsy data stored on Wright-Patterson's classified computer systems.

In 1993, *Dateline NBC* broke its silence on this controversial issue, and Susan Adams, who had been the producer of that particular segment of the program, expressed considerable surprise at the incredible response that NBC had received following the airing of the episode at issue. Surprising absolutely no one at all, the hacker, Adams explained, wished to remain wholly anonymous, primarily because much of his material had allegedly been acquired under illegal circumstances, via delving into classified American government and military computerized files.

Adams was at pains to point out that NBC's lawyers had scrutinized the relevant piece "with a fine-tooth comb" and had duly become convinced of the complete legitimacy of the acquired material. Moreover, Adams continued, and conceded, as the hacker was technically committing nothing less than an outright felony, there was absolutely no way whatsoever that his identity could ever be revealed to the general public or the media. "The hacker is aware of the interest his apparent UFO data has provoked, but does not wish to respond," Adams clarified.[2]

The collective U.S. government, perhaps astutely realizing that trying to uncover the hacker's real identity might very well require bringing a lawsuit against a body as powerful as NBC—which would surely have guaranteed huge, and very unwelcome publicity for the curious affair—chose to do absolutely nothing at all (aside from brood and seethe in silence, one is most strongly inclined to suspect).

CONFIDENTIAL

While such alien information acquired via computer hacking has been overwhelmingly scoffed at by the skeptics and debunkers, it received a major boost in the mid to late 1990s from a wholly unexpected source: a brilliant teenage computer hacker operating from the comfort of his parents' home in the city of Cardiff, Wales, United Kingdom. As an experienced hacker of thousands of computer systems (including many in NASA and the United States Department of Defense), Matthew Bevan made the decision back in 1994 to uncover the UFO secrets of Wright-Patterson Air Force Base. Stressing that Wright-Patterson was "a very, very easy computer system to get into," he was utterly amazed to uncover astonishing information relating to a super-secret project to design and build a truly extraordinary flying machine of UFO-like proportions.[3]

"The files," Bevan told me, in somewhat guarded and hesitant tones in a personal interview, "very clearly referred to a working prototype of an anti-gravity vehicle that utilized a heavy element to power it. This wasn't a normal aircraft; it was very small, split level, with a reactor at the bottom and room for the crew at the top."[4] Might this, perhaps, have been prime evidence of the highly advanced, alien-derived technology that Steven Greer's Disclosure Project maintains is being hidden from us? Quite possibly, yes.

Having accessed and carefully digested the fantastic information, Bevan duly exited the Wright-Patterson computer banks and began to doggedly search everywhere for the alien answers that he sought, including the less-than-secure computer systems of NASA itself. Bevan got into the systems, carefully read the files, and then made good his escape, all without any form of detection whatsoever. Everything was looking pretty good for the young computer hacker.

Or so Bevan had assumed was the case. History, however, has shown that Bevan's initial assumptions were very wide of the mark. For approximately two years there was nothing but overwhelming silence. Then, on a particular morning in 1996, everything suddenly changed drastically in the life of Matthew Bevan. At the time when things began to go distinctly awry, he was working for an insurance company in Cardiff, and on the day in question he was summoned down to the managing director's office. On entering the room, he was confronted by a group of men in suits who seemed to practically ooze intimidation. Bevan recalled what happened next: "One of the men outstretched his hand and I shook it. 'Matthew Bevan?' he asked. 'Yes,' I replied. He said: 'My name is Detective Sergeant Simon Janes of Scotland Yard's Computer Crimes Unit, and I'm placing you under arrest for hacking NASA and Wright-Patterson Air Force Base.'"[5] Bevan was in deep trouble.

On being taken to Cardiff Central Police Station, the line of questioning became decidedly curious and worthy of an episode of *The X-Files*: "What does the term *Hangar 18* mean to you?" Bevan was immediately asked, in stern and intimidating tones. "That's a hoarding place for alien technology," he replied, arms folded, and in a quite matter-of-fact fashion.[6]

Bevan's recollections of that exchange were more than eye-opening:

> Throughout the interview, they kept coming back to Hangar 18: Did I see anything on the Wright-Patterson and the NASA computers? Did I download anything? Well, when they asked me if I saw anything, I said: "Yes, I saw e-mails talking about an anti-gravity propulsion system."[7]

This did not go down too well with Scotland Yard's Computer Crimes Unit. Bevan correctly realized that he was in very hot water with the authorities, and a date was subsequently set for a hearing at London's Bow Street magistrate's court. But it was not just Bevan, his defense, and the prosecution who were present at the trial. A certain Jim Hanson, specifically representing the interests of the U.S. government and NASA, was also in attendance.

There was a curious exchange indeed when Hanson took the stand, as Bevan remembered only too well: "As the hearing continued, the prosecution asked Hanson what the American government thought about my motives regarding my hacking at NASA and at Wright-Patterson," said Bevan. "Hanson replied, 'We now believe that Mr. Bevan had no malicious intentions and that his primary purpose was to uncover information on UFOs and Hangar 18.'" Bevan said, "Well, everyone had a bit of a laugh at that point, even the judge; however, when the prosecution asked: 'Can you confirm if *Hangar 18* exists or if it's a myth?' Hanson said: 'I can neither confirm nor deny as I'm not in possession of that information.'"[8]

The final outcome of the affair was that the case against Bevan completely collapsed. The magistrate overseeing the matter stated in no uncertain terms that a jail sentence was utterly out of the question, and that any financial punishment he might be able to impose upon Bevan would be meager in the extreme. Coupled with the fact that neither NASA nor the American government as a whole was willing to divulge any information concerning the contents of the material on the Wright-Patterson computers to the British court, the cost of prosecuting the case was perceived as being as high as $10,000 a day, so the prosecution grudgingly elected to offer zero evidence. Bevan was, much to his satisfaction, a free man. And he was a very lucky man too.

The total failure to secure a successful conviction—even a minor one—in the Bevan affair both frustrated and angered NASA and the

U.S. government. And the very fact that American authorities flatly refused to reveal to the British court the precise nature and content of the files and the data that Bevan had hacked into made it all but inevitable that the judge would have no choice but to completely dismiss the case. The U.S. government, however, was most definitely not prepared to make that same fatal mistake again when Gary McKinnon hacked into certain NASA computer systems—as will now become apparent.

TOP SECRET

Gary McKinnon's online activity in pursuit of the startling truth about UFOs and alien life forms was destined to follow a very different, and extremely hazardous path from that of most UFO researchers. It was a path that closely paralleled the actions of Welshman Matthew Bevan back in the mid 1990s. But, whereas Bevan had walked away from the British court system as a totally free and a very relieved young man, things have not been anywhere quite so easy and simple for McKinnon. When McKinnon first decided to hack into the computer systems of NASA and the U.S. government for secret UFO data, said British journalist Jon Ronson, he did so from within the house of his girlfriend's aunt, in Crouch End, London. "Basically, what Gary was looking for—and found time and again—were network administrators within high levels of the U.S. government and military establishments who hadn't bothered to give themselves passwords. That's how he got in."[9]

Having finally successfully unlocked the proverbial door to NASA's secret stash of cosmic data, McKinnon found himself confronted by some truly weird material, including files that referenced what were intriguingly referred to as "Non Terrestrial Officers." In McKinnon's own mind at least, this was potential smoking-gun evidence in support of the theory that, as well as running its very public space-based projects such as the space shuttle, Skylab, and the Apollo missions to the moon, NASA also had a top-secret space-based program in place about which the public and the media knew absolutely nothing at all.

By McKinnon's own admission, most of his early, illegal research was prompted by the findings and the revelations of Steven Greer's Disclosure Project, and by the successful, albeit equally illegal actions of Matthew Bevan. In addition, McKinnon told Matthew Williams, a good friend of both me and Matthew Bevan, and a former special investigator with the British government's Customs & Excise agency, that his discovery that NASA and elements of the U.S. government were possibly sitting on top of amazing, secret technologies that had the ability to revolutionize—for the better—the whole planet, made him angry in the extreme. If McKinnon should one day prove to be correct, then he would have every right to be angry—as would all of us.

On the specific matter of Donna Hare's startling disclosures about her firsthand knowledge of NASA's censored UFO photographs, McKinnon has confirmed that it was indeed her out-of-this world revelations that directly set him upon the quest that ultimately landed him in such hot water with American authorities. As a result of being spurred on by Hare's account of her days at the Johnson Space Center, McKinnon successfully accessed that installation with what turned out to be surprising ease. And he duly stumbled upon some truly memorable imagery.

McKinnon explained to Matthew Williams:

> This was my best and worst moment in it all, and I still think back with anger because of the way things went. What [Donna Hare] said was there, was there. I wanted to see the images, and I had to see them. I had to know. But, transferring those files at that size would have taken days; so I had an idea. I would look at it on their screen. I did it by taking graphical control of their desktop and turning the color right down, so that it could transfer to my PC quickly. I saw probably about two-thirds of this picture, and I saw what looked like the Earth's hemisphere with clouds. But then the structure started to appear and it started to reveal the body of what at first looked like a satellite. Then, as it revealed more, I realized that this looked very different and I was onto something. There didn't appear to be any seams or rivets, and no telemetry, no aerials. Just then, I saw the mouse move on the screen and it went down to the lower part of the screen, and next chose the 'Disconnect' command, and that was it: that was me out of NASA. Hats off to NASA: They did close

off my method of entry in practically no time at all, in nearly all of their systems. It was a horrible moment, though, because it was "eureka," and then instantly I got caught.[10]

Williams wanted to know if McKinnon considered his actions to be morally wrong or illegal, or if he even realized that this is how many people would perceive his actions. In response, McKinnon said to Williams that although he knew full well that what he was doing violated the law to a significant degree, he nevertheless felt that his actions were justified as a result of his conclusion that "just because it was illegal doesn't make it wrong." In other words, therefore, in his view, what McKinnon was doing, he added to Williams, was "for the greater public good."[11]

As well as hacking into NASA in search of secret UFO photographs and data, McKinnon went on to do something else that was highly reckless, and that ultimately came back to haunt him. After the U.S. invasion of Afghanistan, he began leaving politically charged and inflammatory messages on U.S. computer terminals, accusing the American government of being directly responsible for 9-11 and of perpetrating an inside job on the American population and the rest of the world, as a means to further the so-called War on Terror.

The very fact that, regardless of his personal motivations for hacking NASA and other elements of U.S. officialdom, McKinnon's actions and comments relative to the 9-11 controversy most assuredly were not undertaken by accident, it is very little wonder that he went on to attract the large-scale attention of officialdom, on both sides of the Atlantic, that was destined to result in his arrest. At first it seemed that McKinnon was all set to get off relatively lightly for intruding into the classified computer systems of NASA and the American government, just as had been the case with Matthew Bevan back in the 1990s.

After having been identified, and subsequently arrested, in 2002 by the U.K. National Hi-Tech Crime Unit, under certain legislation contained in the British Government's Computer Misuse Act, things actually didn't look too bad for McKinnon, surprisingly enough. At the time in question, the terms of the act carried an absolute maximum sentence of only six months' jail time; good behavior on McKinnon's part while behind bars might even have lessened that time period to practically nothing.

In the immediate years after his arrest, not much happened to suggest that things might get any worse for McKinnon. But eventually, matters began to change alarmingly for the hacker, and particularly so when the British Government implemented a new extradition treaty with the U.S. Government that effectively meant American authorities could now extradite a British citizen, in certain specific circumstances, and, somewhat controversially, did not even need to provide any contestable evidence (this was one of the central issues that ultimately led to the collapse of the prosecution team's case in the 1990s trial against Matthew Bevan). American authorities could simply request that British law enforcement personnel hand McKinnon over for a flight to the United States and a behind-closed-doors, media-barred trial in the—ahem—Land of the Free. The timing could not have been any worse for the hacker, particularly when stories surfaced to the effect that he might very well end up serving a substantial period of time in a U.S. maximum-security prison—perhaps as much as 70 years. What had started out as a bit of an *X-Files*-type adventure for McKinnon was rapidly mutating into his worst nightmare.

The U.S. government brought the hammer down hard on the UFO-obsessed McKinnon, accusing him of hacking close to 100 computers in a two-year period from 2000 to 2002, while using the hacking nickname of *Solo*. Those computers, asserted the government, belonged to NASA, the Department of Defense, and the U.S. Air Force, Navy, and Army. The U.S. government's position on McKinnon and his actions was firmly spelled out by a senior military officer at the Pentagon, who told Britain's *Sunday Telegraph* newspaper in no uncertain terms that "we suffered serious damage. This was not some harmless incident. He did very serious and deliberate damage to military and NASA computers and left silly and anti-America messages. All the evidence was that someone was staging a very serious attack on U.S. computer systems."

To this day, McKinnon vehemently disputes the serious charges of causing damage—large or small—to the computer systems that he readily admits having penetrated. On this particular matter, and in no uncertain terms, he told Matthew Williams:

> [The U.S. Government] has kind of redefined "damage." On one level, they have said "impairing the machine's ability to perform its normal function," which is rubbish, because all I did was log on and install my remote control software, which doesn't actually inhibit the machine's ability to function in any way. Then, they go on to say "damage by alteration of data," which refers to the act of installing the remote control software. But, I haven't actually damaged their data in any way by doing so. It is an addition to the machine; not a damage to their data.[12]

Whether computer systems were actually damaged or were merely added to is a moot point; McKinnon had very good reasons to regret his actions: in July 2006, things got much worse—if such a thing was even possible. A ruling was finally made to allow for McKinnon to be extradited to the United States, where he would duly stand trial for his extraterrestrial actions. The court system, as it typically does, lumbered along at an infinitely slow pace, however, and in February 2007, McKinnon's lawyers strongly argued against the ruling in an appeal to the High Court in London. The court was having none of it, however: on April 3, 2007, the appeal was thrown out. Three months later, on July 30, the British government's House of Lords agreed to consider McKinnon's appeal. Again, it was all to no avail: The official verdict was that McKinnon was still ripe for extradition to the United States, and there was now no turning back; he would have to face up to his actions and take his punishment, whatever that might be. Fortunately for McKinnon, his lawyer was granted two weeks to appeal the decision with the European Court of Human Rights before extradition. Yet again, however, McKinnon's appeal was rejected. The clock on McKinnon was ticking ever faster and ever louder. It was time for his legal team to try a radically different approach.

In August 2008, McKinnon was diagnosed in England by Cambridge University professor Simon Baron-Cohen with Asperger's Syndrome, a condition in which the sufferer has impaired and limited social interaction, and a deep preoccupation with one particular issue—which summed up McKinnon's situation to a tee when it came to his lifestyle, and his interest in UFOs and hacking. It was argued that, as a result of this diagnosis, McKinnon was, psychologically and mentally, simply not in a position to stand trial. Despite all that had happened, he was still managing to precariously cling on to his freedom.

Then, on October 27, 2009, there was a further ray of light on the horizon when the Home Secretary of Britain's then Labor Government, Alan Johnson, stated that he was going to do his absolute utmost to bring to a halt the proceedings to have McKinnon extradited, and allow for a comprehensive study of McKinnon's medical condition to proceed just as soon as was bureaucratically possible. The hope that this would help McKinnon's case was short-lived, however. In a letter sent to McKinnon's solicitor, dated November 26, 2009, a representative of the British government's Home Office stated that the extradition of McKinnon would not be a violation of his right as a British citizen, and that, as a result, the extradition would not be placed on hold or overturned after all.

And there are yet further developments in the seemingly never-ending affair (or perhaps *soap opera* would be a more accurate term) of Gary McKinnon. In May 2010, Britain's Labor Party was voted out of power by the British people, and a coalition government, led by the Conservative Party's David Cameron, took control. Following the election, it was announced that Cameron's new Home Secretary, Theresa May, was willing to address the decisions that had previously been made under Gordon Brown's near-Orwellian Labor Government with respect to McKinnon and his reckless actions, and determine if extradition was really justified, or if McKinnon should be tried for his activities within the borders of the British Isles, where even NASA readily agreed his crimes had all been committed.

Then on June 13, 2010, the British media said that Prime Minister Cameron had been given the go-ahead to finally take decisive action and save McKinnon from extradition to the United States to face trial, and, potentially, decades behind bars in an American jail. Theresa May had briefed the prime minister on McKinnon's case so that he could halt extradition proceedings, opening up the possibility of McKinnon being tried for his offence in the U.K.—and also of McKinnon serving prison time in the U.K. too, if such was warranted by the verdict of the court.

McKinnon's mother was desperately trying to stay positive about this latest episode:

> We're all very nervous at the moment and hoping for good news and that Gary will soon have his life back again.[13]

The Home Office, meanwhile, was said to be "currently considering representations from Mr. McKinnon's legal team," while a Home Office spokeswoman added, when questioned about this reported consideration:

> It is not appropriate to speculate further at this stage.[14]

And, at the time that I type these words, that is where matters stand: McKinnon remains free (albeit still with a veritable sword of Damocles hanging precariously over his stressed-out head), NASA and certain senior elements of the U.S. government and military continue to seethe, the issue of the censored UFO pictures that McKinnon said he found on NASA's computers continues to tantalize UFO researchers, investigators, and even the media, and the ultimate outcome is still nowhere in sight.

But, if the case of Gary McKinnon does eventually go to trial, then we may be in for some interesting developments and revelations—if the trial does not take place behind closed doors, with the media barred from covering it (in the name of national security, of course). And here's why we may see a few surprises: While I was in the process of undertaking research for this chapter, I had several extended telephone conversations with personnel from within NASA's Public Affairs office. This, in turn, led me to speak with a NASA employee working in the field of legal issues as they relate to the affairs of the space agency, and who had not only followed the McKinnon saga closely, but who also was well-versed in the scope of the case, its complexities, and its implications—not just for McKinnon, but for anyone else who might be unwise enough to think about someday following a similar path. The man also expressed a few sympathies for McKinnon's plight, as he saw it.

I was told that if the U.S. government does press ahead with its case against Gary McKinnon, and he is subsequently brought to the United

States to stand trial, then the government, and NASA, may not get the smooth ride they may be anticipating. When I asked what he meant by this, the man replied that everyone involved in the affair—McKinnon's legal team, the government, and even NASA—all agreed that, regardless of the right or wrong of McKinnon's actions, they were pretty much exclusively spurred on by the testimony of Donna Hare, who worked for a subcontractor to NASA.

Had McKinnon not been aware of Hare's testimony about allegedly censored UFO photos held by NASA at its Johnson Space Center, opined the man, McKinnon's legal team might be in a position to legitimately argue that he would never have embarked upon the hacking spree that has placed him in his current position.

With that in mind, there might even be grounds, added the man, to push Hare to testify before the court about her revelations (which, remember, had extended to the specific location where the censoring was said to be taking place). And, the man further speculated, it might be suggested that McKinnon was only guilty of hacking because a NASA subcontractor was guilty of spilling the beans on classified UFO-related data held at the Johnson Space Center in the first place. As the man asked me, where does the buck stop: with McKinnon for hacking, with Hare for revealing the story and the location, or with NASA for originally and (perhaps even illegally) hiding the truth?

It was not entirely out of the question, the man suggested to me, that a very good legal argument could be made that because someone subcontracted to NASA had told people where such activity was specifically afoot, it was hardly the sole fault of McKinnon, taking into account his mental condition and his diagnosis of Asperger's, that he then decided to go looking for it.

The man admitted that such a ploy would likely not impress the prosecution team at all, and might even result in warnings from the judge, but what it could conceivably result in would be a call for Hare to take the stand and offer testimony about her revelations back in 2001 that set McKinnon on the alien trail. And this would be something that neither NASA nor the U.S. government would want, and, if it did go ahead, might actually result in U.S. officialdom quickly and quietly backing off from prosecution, lest such prosecution of McKinnon only serve to risk opening up even more ufological doors.

The man further offered suggestions that, if McKinnon does one day go to trial, his legal team should explain to the judge that in at least one case of a very similar nature the U.S. government elected to take absolutely no action at all. On this matter, he referred me specifically to the case of the computer hacker whose actions were highlighted on NBC's *Dateline* show in 1993, and Matthew Bevan. And, as the man noted, Bevan walked totally free from a British court, after the U.S. government and NASA refused to provide any hard evidence of his actions, or any data that reflected the nature of the files into which Bevan hacked.

In other words, the man was saying to me, there were at least a couple of relatively recent precedents in which computer hackers seeking UFO material—and, it would appear, UFO material *only*—were allowed to walk free, or were not even arrested, charged, or brought to trial at all. And with those important precedents in mind, any judge might be persuaded that a serious prosecution of McKinnon might not be a successful, or even viable option. Time may tell if the views and thoughts of the man prove to be correct or not.

The final words, for now at least, go to Gary McKinnon, who, reflecting on the dire situation in which he still finds himself immersed, told Matthew Williams:

> My dad would say: "It's your own stupid fault, kid!"[15]

Conclusion
Past, Present, and Future

Having now digested a wealth of data, testimony, observations, whistleblower words, and pages of official and unofficial documentation, what can be said about NASA and the claims of high-level conspiracy related to the moon landings, the Face on Mars, censored photographs, UFOs, alien life, crashed flying saucers, alien abductions, Contactee cases, and much more?

Despite the fact that there is an ever-growing group out there that loudly and confidently asserts the Apollo moon landings were nothing but prime examples of ingenious Hollywood-style fakery, the truth does seem to rest with NASA: The landings, it would appear, did go ahead just as NASA has always claimed.

In addition, it seems fairly safe to conclude that we can put to rest the many and varied conspiracy theories that were spawned in the immediate wakes of the tragic destruction of the space shuttles *Challenger* in 1986 and *Columbia* in 2003. However, the fact that the FBI has chosen to continue to withhold certain files and documents on its investigation of the *Challenger* explosion in the name of national security nearly a quarter of a century later is something that will likely keep the flame of conspiracy alight on this matter for some time.

On certain other out-of-this-world controversies, however, NASA might not be so in the clear as it would undoubtedly prefer to be: the Kecksburg, Pennsylvania, affair of 1965, the Bolivian event of 1978, the Roswell-connected words of Apollo astronaut Dr. Edgar Mitchell, the issue of lethal alien viruses, and the strange saga of NASA and the chupacabra, some would argue, are all prime evidence that NASA may know far more about UFOs and extraterrestrial activity than it cares to publicly admit.

When it comes to the still-thriving controversy surrounding the Face on Mars, the late Mac Tonnies made a very strong, logical case to the effect that dismissing the face as a mere trick of the light could prove to be a major disaster, scientifically, historically, and culturally. It remains to be seen whether NASA is actually guilty of hiding hard evidence that the face is an artificial construct, or merely prefers to play down the whole matter because it has become tired of dealing with accusations that it is sitting on top of secret proof that intelligent, long-extinct Martians constructed the face countless millennia ago.

Turning our attention toward the alternative actions of computer hacker Gary McKinnon, the work of the Disclosure Project, the revelations of Donna Hare and Karl Wolf, and the issue of NASA's censored photographs of a UFO-connected nature, the probability that much more still remains to surface on these matters seems to be extremely likely indeed, as does the probability that where there is a great deal of extraterrestrial smoke, there is almost certainly an equally large amount of extraterrestrial fire. And, if McKinnon does go to trial, perhaps we will finally get to see some of that fire.

The illuminating and amazing experiences of P.T. McGavin and Sharon (as they relate to the so-called Contactee controversy and the alien abduction epidemic) strongly suggest that NASA knows a great deal about both matters, even if some of their conclusions about ancient human civilizations and demonic entities do not sit too well with those who solely look to the skies and the stars for the answers to the mystery of unidentified flying objects.

But, moving on from the past, the most important question that currently faces us now is: What may the future bring our way? Are we ever likely to know the full and unexpurgated stories behind the many NASA conspiracies that have been detailed within the pages of this book? We just might. Throughout the course of the last few years, numerous rumors have surfaced across the Internet, and within the UFO research arena, to the effect that certain senior and powerful elements within NASA and the governments and infrastructure of the United States, Great Britain, Russia, and other leading powers have secretly known for decades that extraterrestrials are visiting the Earth.

According to many players in the UFO research arena, en masse disclosure of this undoubtedly paradigm-shattering news will very soon be forthcoming. In addition, swirling rumors suggest that performing integral roles in the worldwide revelations of the alien kind will be both NASA and the Vatican. Make no mistake: The Vatican has a very deep and vested interest in the issue of whether or not some form of alien life exists within the vast expanses of the Universe. Moreover, the Vatican's own representatives and staff have chosen—some might even say ordered—to be curiously and significantly vocal on this matter in recent years. Monsignor Corrado Balducci, a Roman Catholic theologian, before his death in 2008, made a number of intriguing public statements on the issues of alien life, UFOs, and human/extraterrestrial interaction. He was extremely open to the idea that we are not alone in the Universe, that alien intelligence may even be widespread, and that this same unearthly intelligence might very well be responsible for the many UFO sightings that have been occurring for decades.

The year 2008 was notable for another development in the strange saga of the Vatican and all things of extraterrestrial origins. In May of that year, the Vatican announced that a firm acceptance in the existence of alien life, whether right now or still light-years away from Earth, did not clash in the slightest with Christian belief systems or with the Vatican's long-term teachings. Indeed, a prominent article that went by the memorable title of "Aliens are my Brother" was splashed across the pages of the Vatican's own journal, *L'Osservatore Romano*, in precisely the same month that the Vatican made its groundbreaking announcement. The article highlighted the words of one Father Gabriel Funes, who offered an opinion that astrobiology should be considered a valid and worthwhile avenue for investigation and research, and stressed carefully that the possibility that life might exist elsewhere did not conflict with the present-day views of the Church at all.

Also in May 2008, the researcher Thomas Horn noted that a Catholic theologian, Father Malachi Martin, offered an enigmatic reply when asked by host Art Bell on the *Coast to Coast AM* radio show why the Vatican was increasing its investigations into the domain of outer space, and why it was doing so particularly at the Mt. Graham Observatory in southeastern Arizona. Martin's reply was as carefully worded as it was diplomatic—and strange. It was highly provocative too:

> ...the mentality amongst those at the highest levels of Vatican administration and geopolitics, know that, now, knowledge of what's going on in space, and what's approaching us, could be of great import in the next five years, 10 years.[1]

Then, in November 2009, as NASA itself announced in a widely circulated press release:

> This past week in Rome as part of the International Year of Astronomy, the Pontifical Academy of Sciences hosted a Study Week on Astrobiology.... Their discussion ranges from what it would mean to the Church if alien life were found, to whether or not science needs religion.[2]

This final sentence from NASA's November 2009 press release is highly significant, because it closely echoes the words and recommendations of the Brookings report of 1960, "Proposed Studies on the Implications of Peaceful Space Activities for Human Affairs" (cited in Chapter 1 of this book). On the matter of a revelation to the public that alien life had been confirmed, Brookings advised NASA 50 years ago that an individual's reaction to such contact with a nonhuman intelligence "would in part depend on his cultural, religious, and social background, as well as on the actions of those he considered authorities and leaders, and their behavior, in turn would in part depend on their cultural, social, and religious environment."[3]

In other words, from its earliest formative years, NASA recognized that religion would very likely play an integral role in determining the outcome of any sort of large-scale disclosure to the public that extraterrestrial life was an undeniable reality. Time may soon tell whether

NASA's relationship to the Vatican and its increasing role in commenting on matters pertaining to life elsewhere in the Universe are integral parts of a greater design to slowly prepare the public for the day when the alien truth is finally revealed (in accordance with the suggestions, guidelines, and ideas of the Brookings Institution, laid down before NASA at the dawning of the 1960s).

The NASA conspiracies, we would do well to consider, may very soon be conspiracies no more.

Notes

Chapter 1

1. FBI files, 1957.
2. Ibid.
3. Redfern, interviews with Mac Tonnies, 2004.
4. Ibid.

Chapter 2

1. "Project Mercury Overview," NASA.gov.
2. Cooper, *Leap of Faith*.
3. Spence, Edwards Air Force Base.
4. Gairy, statement to the United Nations.
5. Ibid.
6. Ibid.
7. Department of State files on Gordon Cooper, 1978.
8. "Gordon Cooper on Soviet Domination," *Omni Magazine*.
9. Department of State files on Gordon Cooper.
10. Slayton, *Deke!*
11. Ibid.
12. Ibid.
13. FBI, 1952.
14. "Gemini Overview," NASA.gov.
15. National Capital Area Skeptics, "Scientific Study."
16. FBI files, 1965.
17. Ibid.
18. Ibid.

Chapter 3

1. Gordon, "Kecksburg Incident."
2. KDKA Radio broadcast.
3. Gordon, "Kecksburg Incident."
4. Kean, "The Cosmos 96 Question."
5. Kean, "The Conclusion of the NASA Lawsuit."

6. Podesta, John, statement.

Chapter 4

1. De Forest, Dr. Lee, statement.
2. Mailer, *Of a Fire on the Moon*.
3. Streuli, "NASA Hires Writer."
4. Brooks, *Chariots for Apollo*.
5. Ibid.
6. Mechanic, "Polis Report."
7. Clinton, *My Life*.
8. White House, "In Event of Moon Disaster."

Chapter 5

1. Redfern, interview with John, 2006.
2. Ibid.
3. Ibid.
4. Ibid.
5. Ibid.

Chapter 6

1. Covert, "A History of Fort Detrick."
2. *New York Times*, Letter from Joshua Lederberg.
3. Lederberg, "Engineering Viruses."
4. Mullen, "Alien Infection."
5. Lovgren, "Far-Out Theory."

Chapter 7

1. Department of Defense UFO files, 1974.
2. FBI files on UFOs, 1957.
3. Department of Defense UFO files, 1974.
4. Ibid.
5. Ibid.

Chapter 8

1. Jauregui, "Occupant Encounter."
2. Redfern, interview with P.T. McGavin, 2003.
3. Ibid.
4. Ibid.
5. Ibid.
6. Ibid.
7. Ibid.
8. Ibid.
9. Ibid.
10. Redfern, interviews with Mac Tonnies, February 2009.
11. Ibid.
12. Ministry of Defense UFO file, 1965.
13. Redfern, interview with Nick Pope, 1998.
14. Ibid.
15. Ibid.
16. Pope, *Operation Thunder Child*.
17. Redfern, interview with Nick Pope, 1998.
18. Ibid.

Chapter 9

1. Redfern, interview with Mac Tonnies, March 2004.
2. Ibid.
3. Ibid.
4. Ibid.
5. Ibid.
6. Ibid.

7. Ibid.
8. Ibid.
9. Ibid.
10. Ibid.
11. Ibid.
12. Ibid.
13. Ibid.
14. Redfern, interview with Mac Tonnies, September 2006.
15. Ibid.
16. Redfern, interview with Mac Tonnies, March 2004.
17. Ibid.
18. Ibid.
19. Ibid.
20. Ibid.
21. Ibid.
22. Redfern, interview with Mac Tonnies, June 2005.

Chapter 10

1. CAUS, *Just Cause*, May 1978.
2. Ibid.
3. "Report of Fallen Space Object," Department of State, 1978.
4. "Reports Conflict," Central Intelligence Agency, 1978.
5. Ibid.
6. Ibid.
7. Ibid.
8. Ibid.
9. Ibid.
10. U.S. Defense Attaché Office, Bolivia, 1978.
11. CAUS, *Just Cause*, August 1978.
12. Ibid.
13. Fawcett, *Clear Intent*.
14. Stringfield, *The UFO Crash/Retrieval Syndrome*.
15. Defense Intelligence Agency report, August 17, 1979.
16. Defense Intelligence Agency report, August 21, 1979.
17. "Satellite Accidents," Home Office.
18. Ibid.
19. Redfern, interview with Peter Jeffries, 2007.

Chapter 11

1. Barnett, "U.S. Planned."
2. "Carl Sagan, FBI File."
3. Ibid..
4. Ibid.
5. Ibid.
6. Ibid.
7. Ibid.
8. Ibid.
9. "Report of the Presidential Commission," NASA.
10. Redfern, interview with Martin Black, 2003.
11. Stringfield, *UFO Crash/Retrievals*.
12. Dolan, "Of Astronauts and Aliens."
13. "Space Shuttle Challenger Explosion," FBI.
14. Ibid.
15. Ibid.
16. Ibid.
17. Ibid.

18. Ibid.
19. Ibid.
20. Ibid.
21. Ibid.
22. Ibid.
23. Ibid.
24. Ibid.
25. Ibid.
26. Ibid.
27. Ibid.
28. Mendenhall, "Amid sympathy."
29. Ibid.

Chapter 12

1. Redfern, interview with Sharon, 2008.
2. Ibid.
3. Ibid.
4. Redfern, 2010.
5. Ibid.
6. Ibid.

Chapter 13

1. "Bat Man Mystery," *Houston Chronicle*.
2. Gerhard, *Monsters of Texas*.
3. Ibid.
4. Ibid.
5. Redfern, interview with Desiree Shaw, 2004.
6. Drake, *Is Anyone Out There?*
7. Ibid.
8. Redfern, interview with Bruce Weaver, 2009.

Chapter 14

1. Department of State files on UFOs in Brazil, 1986.
2. Ibid.
3. CIA files on UFO encounters, 1989.
4. Ibid.
5. Ibid.
6. Ibid.
7. Ibid.
8. Ibid.
9. "Somaliland President," CIA.

Chapter 15

1. "RAAF Captures Flying Saucer," *Roswell Daily Record*.
2. Berlitz, *The Roswell Incident*.
3. Ibid.
4. Moore, William L., *The Roswell Investigation*.
5. Ibid.
6. Moore, William L., "Crashed Saucers."
7. Randle and Schmitt, *UFO Crash at Roswell*, and *The Truth About the UFO Crash at Roswell*.
8. Rhodes, "UFOs."
9. Earls, "Yes, Aliens Really Are Out There."
10. Moore, Waveney Ann, "Astronaut."
11. Margerrison, *Kerrang Radio*.
12. "Apollo 14 Astronaut Claims," *Daily Mail*.

13. "Apollo 14 Astronaut Edgar Mitchell Responds.," *Shape Shifting*.
14. "Klotz, "Apollo Astronaut."

Chapter 16

1. NASA files on Joseph Perry, 1960.
2. FBI files on Joseph Perry, 1960.
3. Wilhelm, "Grand Blanc Man."
4. Ibid.
5. NASA files on Joseph Perry, 1960.

Chapter 17

1. Goldwater, letter to Shlomo Arnon, 1975.
2. "Are Your Secrets Safe?" *Dateline NBC*.
3. Redfern, interview with Matthew Bevan, 1998.
4. Ibid.
5. Ibid.
6. Ibid.
7. Ibid.
8. Ibid.
9. Ronson, "Jon Ronson Meeets Hacker Gary McKinnon."
10. Williams, "A Close Encounter."
11. Ibid.
12. Sherwell, "Hacker Gary McKinnon."
13. Williams, "A Close Encounter."
14. Borland, "David Cameron."
15. Williams, "A Close Encounter."

Conclusion

1. Horn, "Is the Vatican Easing Humanity."
2. Scalice, "Vatican Hosts."
3. Michael, "Proposed Studies."

Bibliography

Note: Date of access for each Website given is August, 2010.

Achenbach, Joel. "NASA Budget for 2011 Eliminates Funds for Manned Lunar Missions." *Washington Post*, February 1, 2010.

Adams, Mike. "The Complete Lee de Forest," *www.leedeforest.org*, 2003.

"Air Force had Plans to Nuke Moon." *www.space.com/news/spacehistory/nuke_moon_000514.html*, May 14, 2000.

"Apollo 11 and Nixon." National Archives and Records Administration. *www.archives.gov/exhibits/american_originals/apollo11.html*, March 1996.

"Apollo 14 Astronaut Claims Aliens HAVE Made Contact—But it Has Been Covered up for 60 Years." *Daily Mail*, July 24, 2008. *www.dailymail.co.uk/sciencetech/article-1037471/Apollo-14-astronaut-claims-aliens-HAVE-contact--covered-60-years.html*.

"Apollo 14 Astronaut Edgar Mitchell Responds..." *Shape Shifting With Lisa Bonnice Radio Show*. *www.blogtalkradio.com/shapeshifting/2008/07/24/apollo-14-astronaut-edgar-mitchell-responds*, July 24, 2008.

"Are Your Secrets Safe?" *Dateline NBC*, October 27, 1992.

"Army Ropes off Area: Unidentified Flying Object Falls near Kecksburg." *Tribune-Review*, December 10, 1965.

"Astronaut Gordon Cooper Witnesses UFO Landing at Edwards AFB." *www.ufoevidence.org/cases/case357.htm.*

"Average Hacker Skills Shut Down US Defense Systems." *Computer Weekly. www.computerweekly.com/Articles/2002/11/21/191125/39Average-hacker39-skills-shut-down-US-defence.htm*, November 21, 2002.

Barnett, Anthony. "U.S. Planned One Big Nuclear Blast for Mankind." *Observer*, May 14, 2000.

Baron, Ronald Thomas. "An Apollo Report, September 1965–November 1966." *www.hq.nasa.gov/office/pao/History/Apollo204/barron.html*, February 3, 2003.

"Bat Man Mystery." *Houston Chronicle*, June 20, 1953.

Batty, David. "New Medical Evidence Could Stop Hacker Gary McKinnon's Extradition." *www.guardian.co.uk/world/2009/oct/26/garry-mckinnon-extradition-alan-johnson*, October 26, 2009.

Berlitz, Charles, and William L. Moore. *The Roswell Incident*. London: Granada Publishing, Ltd., 1980.

"Best Evidence: Donna Hare Speaks the Truth About NASA, The." *http://thebestevidence.blogspot.com/2009/06/donna-hare-speaks-truth-about-nasa.html*, June, 2006.

Billingham, J., and B.M. Oliver. "Project Cyclops: A Design Study of a System for Detecting Extraterrestrial Life." NASA, 1972.

Bilstein, Roger E. "Orders of Magnitude: A History of the NACA and NASA, 1915–1990." NASA, 1989.

"Biographical Data: Deke Slayton." *www11.jsc.nasa.gov/Bios/htmlbios/slayton.html*, 1993.

"Biographical Data: Leroy Gordon Cooper, Jr." *www.jsc.nasa.gov/Bios/htmlbios/cooper-lg.html*, 2004.

Booth, B.J. "The Kecksburg Crash." *http://ufocasebook.com/Kecksburg.html.*

Borland, Ben. "David Cameron Can Save Hacker From US Jail Hell." *www.dailyexpress.co.uk/posts/view/180683/David-Cameron-CAN-save-hacker-from-U-S-Jail-hell/*, June 13, 2010.

Boyle, Alan. "A New View of the Famous Face on Mars." MSNBC. *www.msnbc.msn.com/id/3077691/*, July 25, 2002.

Brian, William L. *Moongate: Suppressed Findings of the U.S. Space Program*. Unknown city: Portland, Or.: Future Science Research Publishing Company, 1982.

Britt, Robert Roy. "Alien Microbe Reported Found in Earth's Atmosphere." *www.space.com/scienceastronomy/planetearth/alien_bacteria_001127.html*, November 27, 2000.

Brooks, Courtney, G., James M. Grimwood, and Loyd S. Swenson. *Chariots for Apollo: A History of Manned Lunar Spacecraft*. NASA Special Publication 4205, NASA History Series, 1979.

Caldwell, Deborah, and Steven Waldman. "A Higher Reason for Columbia Crash?" ABCNews. *http://abcnews.go.com/US/story?id=90879&page=1*, February 6, 2003.

"Carl Sagan, FBI File." FBI, 1983. *http://foia.fbi.gov/foiaindex/sagan_c.htm*.

"Carl Sagan." *http://www.carlsagan.com*, 2010.

CAUS. "Bolivian Documents Released by State Department: Mystery Deepens." *Just Cause*, August 1978.

———. "U.S. Agencies Scratch Heads Over Bolivian Incident." *Just Cause*, June 1978.

———. "Crashed UFO in Bolivia?" *Just Cause*, May 1978.

Chaikin, Andrew. "White House Tapes Shed Light on JFK Space Race Legend." *www.space.com/news/kennedy_tapes_010822.html*.

CIA files on UFO encounters in the former Soviet Union. (Declassified under FOIA.) 1989.

Clinton, Bill. *My Life*. New York: Knopf, 2004.

"Coalition for Freedom of Information." *www.freedomofinfo.org/freedom.html*.

Cocconi, Giuseppe, and Philip Morrison. "Searching for Interstellar Communications." *Nature*, September 19, 1959.

"Commenting on Recent Disclosure with Kerrang Radio." *Fox News*, July 25, 2008.

Committee for Skeptical Inquiry. "Buzz Aldrin Punches Moon-Landing Conspiracy Theorist." *www.csicop.org/news/show/buzz_aldrin_punches_moon-landing_conspiracy_theorist*, October 16, 2002.

"Conspiracy Theories on Space Shuttle Columbia Blame Jews, Israel for Disaster." *www.adl.org/PresRele/IslME_62/4245_62.htm*, February, 18, 2003.

Cooper, Gordon, with Bruce Henderson. *Leap of Faith: An Astronaut's Journey into the Unknown*. New York: Harper-Collins, 2000.

Cooper, Gordon L. Letter to Mission of Grenada to the United Nations. November 9, 1978.

Cosmos: A Personal Voyage. PBS/KCET, 1980.

Cosnette, Dave. "The Apollo Hoax." *www.ufos-aliens.co.uk/cosmicapollo.html*, February 10, 2009.

Covert, Norman. "A History of Fort Detrick, Maryland." *www.detrick.army.mil/cutting_edge/index.cfm*, October 2000.

Creighton, Gordon. "The Curious Utterances of the Vatican's Monsignor Corrado Balducci." Flying Saucer Review 45, No. 4, Winter 2000.

Crichton, Michael. *The Andromeda Strain*. New York: Ballantine Books, 1992.

"Cydonia Quest." *The Occasional Journal*, No. 10. *http://bob-wonderland.supanet.com/journal_11.htm*, March 1, 2006.

Darling, David. "Cyclops, Project." *www.daviddarlinginfo/encyclopedia/C/Cyclops.html*.

———. "'Face' on Mars." *www.daviddarling.info/encyclopedia/F/face.html*.

De Forest, Dr. Lee. Statement given in 1926, cited in the *New York Times*, February 25, 1957.

Defense Intelligence Agency report, August 17, 1979.

Defense Intelligence Agency report, August 21, 1979.

Department of Defense. UFO files, 1974.

Department of State. Files on Gordon Cooper, November 24, 1978.

———. Files on Gordon Cooper, November 15, 1985.

———. Files on UFOs in Brazil, May 1986.

———. Public Affairs, November 23, 2003.

"Disclosure Project, The." *www.disclosureproject.org/*, 2010.

Dolan, Richard M. "Musings on a Secret Space Program." *http://richardthomasblogger.blogspot.com/*, June 6, 2010.

———. "Of Astronauts and Aliens." *www.keyholepublishing.com/Of%20Astronauts%20and%20Aliens.htm*, June 14, 2004.

———. *UFOs & the National Security State: The Cover-Up Exposed, 1973–1991*. Rochester, N.Y.: Keyhole Publishing Company, 2009.

Downes, Jonathan. *Island of Paradise: Chupacabras, UFO Crash Retrievals, and Accelerated Evolution on the Island of Puerto Rico*. Bideford, North Devon, England: CFZ Press, 2008.

Drake, Frank, and Dava Sobel. *Is Anyone Out There?* Concord, Calif.: Delta, 1994.

Dudurich, Ann Saul. "Kecksburg UFO Debate Renewed." *www.PittsburghLive.com*, August 3, 2003.

Earls, John. "Yes, Aliens Really Are Out There, Says the Man on the Moon." *People*, October 25, 1998.

"Ed Mitchell—Apollo 14.com." *www.edmitchellapollo14.com*.

"Edgar Mitchell Interview on Dateline NBC." *www.ufoevidence.org/documents/doc1923.htm*, April 19, 1996.

"Edgar Mitchell on the UFO Cover-Up." *UFO Updates*, 1998.

"Edgar Mitchell UFO Interview on Kerrang Radio." *www.youtube.com/watch?v=RhNdxdveK7c,* July 23, 2008.

Eddington, Colonel Robert. "Report of Fallen Space Object." Telegram sent to Paul H. Boeker, U.S. Embassador to Bolivia. La Paz, Bolivia, May 18, 1978.

"Ellison S. Onizuka." *www.jsc.nasa.gov/Bios/htmlbios/onizuka.html*, January 2007.

Fawcett, Lawrence, and Barry J. Greenwood. *Clear Intent: The Government Cover-Up of the UFO Experience*. Upper Saddle River, N.J.: Prentice-Hall, Inc., 1984.

FBI files on Joseph Perry. (Declassified under FOIA). 1960.

FBI files on UFOs. (Declassified under FOIA). 1952.

———. (Declassified under FOIA). 1957.

———. (Declassified under FOIA). 1965.

Filer, George A. "Cooper Says Disc Landed, Filmed at Edwards." *Filer's Files*, No. 38, September 26, 2000.

"Fireball a Meteor, Astronomer Explains." *Pittsburgh Post-Gazette*, December 10, 1965.

Flying Saucer Review 28, No. 3, 1983.

"Former Astronaut: Man not Alone in Universe." CNN, April 20, 2009.

"Frank Drake Interview." *www.ufoevidence.org/documents/doc1428.htm*, 2010.

"Free Gary McKinnon." *http://freegary.org.uk/*, 2010.

Friedman, Stanton T., and Don Berliner. *Crash at Corona*. St. Paul, Minn.: Paragon House, 1992.

Friedman, Stanton T., and William L. Moore. "The Roswell Incident: Beginning of the Cosmic Watergate." MUFON Symposium Proceedings, 1981.

Gairy, Sir Eric Matthew. Statement to the United Nations, October 1977.

Garber, Steve. "Apollo 1." *http://history.nasa.gov/Apollo204/*, January 27, 2010.

———. "Apollo 30th Anniversary." *http://history.nasa.gov/ap11ann/introduction.htm*, September 20, 2002.

———. "Sputnik and the Dawn of the Space Age." *http://history.nasa.gov/sputnik/*, October 10, 2007.

Garner, Robert. "LRO Sees Apollo Landing Sites." *www.nasa.gov/mission_pages/LRO/multimedia/lroimages/apollosites.html*, July 17, 2009.

"Gemini Overview." NASA.gov, written 1977. *www-pao.ksc.nasa.gov/history/Gemini/Gemini-overview.htm*.

Gentleman, Amelia, and Robin and McKie. "Red Rain Could Prove That Aliens Have Landed." *Observer*, March 5, 2006.

Gerhard, Ken, and Nick Redfern. *Monsters of Texas*. Bideford, North Devon, England: CFZ Press, 2010.

"Glenn Research Center." *www.grc.nasa.gov*, 2010.

Goldwater, Senator Barry. Letter to Shlomo Arnon, March 28, 1975. Reproduced in Good, Timothy, *Above Top Secret*, London: Sidgwick & Jackson, 1987.

Good, Timothy. *Above Top Secret*. London, UK: Sidgwick & Jackson, 1987.

———. *Alien Contact: Top Secret UFO Files Revealed*. Minneapolis, Minn.: Quill, 1994.

———. *Alien Liaison*. London, UK: Random Century Ltd., 1991.

"Gordon Cooper on Soviet Domination of Space, NASA's Impotence, and UFOs." *Omni Magazine*, March 1980.

Gordon, Stan. "The Kecksburg Incident: An Updated Review." *1st Annual UFO Crash Retrieval Conference*. Wood & Wood Enterprises, 2003.

———. "Stan Gordon's UFO Anomalies Zone: Kecksburg UFO Crash." *www.stangordonufo.com/kecksburg/kecksburg%20home.htm*, 2009.

Gray, Tara. "L. Gordon Cooper, Jr." *http://history.nasa.gov/40thmerc7/cooper.htm*, 2010.

Grayzeck, Dr Ed. "Sputnik 1." *http://nssdc.gsfc.nasa.gov/nmc/spacecraftDisplay.do?id=1957-001B*, July 23, 2010.

"Great Moon Hoax, The." *http://science.nasa.gov/science-news/science-at-nasa/2001/ast23feb_2/*, 2001.

Greer, Steven, M. *Disclosure: Military and Government Witnesses Reveal the Greatest Secrets in Modern History*. Crozet, Va.: Crossing Point, Inc., 2001.

Griggs, Brandon. "Could Moon Landings Have Been Faked? Some Still Think So." *www.cnn.com/2009/TECH/space/07/17/moon.landing.hoax/index.html*, July 17, 2009.

Grimwood, James. M. "Major Events Leading to Project Mercury." *http://history.nasa.gov/SP-4001/p1a.htm*.

Gutheinz, Joseph. "I fear Gary McKinnon will not find justice in America." *www.heraldscotland.com/i-fear-gary-mckinnon-will-not-find-justice-in-america-1.901964*, February 6, 2009.

Hacker, Barton C., and James M. Grimwood. "On the Shoulders of Titans: A History of Project Gemini." NASA, 1977.

Hamilton, Calvin J. "The Great Moon Hoax." *www.solarviews.com/eng/moonhoax.htm*, February 23, 2001.

Hoagland, Richard. "Forbidden Planet...Mars." *www.enterprisemission.com/forbidden-planet.htm*, 2006.

"Holy See: Vatican City State, The." *www.vatican.va/vatican_city_state/index.htm*, 2010.

Horn, Thomas. "Is the Vatican Easing Humanity Toward Alien Disclosure?" *www.americanchronicle.com/articles/view/62208*, May 18, 2008.

Hynek, J. Allen. Speech to the United Nations, November 27, 1978.

"I Saw Structures on the Moon (Karl Wolfe)." *www.ufocasebook.com/moonstructures.html*, (undated).

"In Event of Moon Disaster." White House, July 18, 1969.

"Israel Mourns First Astronaut's Death." *www.cnn.com/2003/WORLD/meast/02/01/shuttle.israel.reax/*, February 1, 2003.

Jad'on, Kelly. "UFO Disclosure 2010: The Vatican's Key Role." *www.basilandspice.com/journal/ufo-disclosure-2010-the-vaticans-key-role.html*, April 13, 2010.

Jauregui, Liria D. "Occupant Encounter in Argentina." *APRO Bulletin* 22, No. 3, 1973.

"John F. Kennedy Moon Speech—Rice Stadium." September 12, 1962. *http://er.jsc.nasa.gov/seh/ricetalk.htm*.

Jones, Eric. M. "Apollo 14, Lunar Surface Journal." *www.hq.nasa.gov/alsj/a14/a14.html*, 1995.

Jorden, William J. "Soviet Fires Earth Satellite into Space." *New York Times*, October 5, 1957.

"Joseph Perry, UFO Photograph 1960, Summary." NASA. March 6, 1960.

"Judgments McKinnon V Government of the United States of America and Another, House of Lords." *www.publications.parliament.uk/pa/ld200708/ldjudgmt/jd080730/mckinn-1.htm*, July 30, 2008.

Kauderer, Amiko. "NASA Human Space Flight." *http://spaceflight.nasa.gov/home/index.html*, June 2, 2010.

Kaysing, Bill. *We Never Went to the Moon: America's Thirty Billion Dollar Swindle*. Palm Springs, Calif.: Desert Publications, 1981.

Kaysing, Wendy L. "A Brief Biography of Bill Kaysing." *http://billkaysing.com/biography.php*, 2006.

KDKA Radio broadcast, December 9, 1965.

Kean, Leslie. "The Conclusion of the NASA Lawsuit Concerning the Kecksburg, PA UFO Case of 1965." *www.freedomofinfo.org/foi/NASA_lawsuit_conclusion.pdf*, November 2009.

———. "The Cosmos 96 Question is Settled Once and for All." *www.freedomofinfo.org/news/cosmos-96.pdf*, October, 2003.

———. "Forty Years of Secrecy: NASA, the Military, and the 1965 Kecksburg Crash." *International UFO Reporter* 30, No. 1.

———. "Project Moon Dust and Operation Blue Fly: The Retrieval of Objects of Unknown Origin." *www.bibliotecapleyades.net/sociopolitica/esp_sociopol_mj12_3k.htm*, 2002.

King, Jon. "U.S. Prosecution of Gary McKinnon 'Spiteful' Says Ex-Top Cop." *www.consciousape.com/news/us-prosecution-of-gary-mckinnon-spiteful-says-ex-top-cop/*, May 10, 2010.

Kirk, Jeremy. "Security Advice from a Wanted Hacker." *PC World*. *www.pcworld.com/article/125584/security_advice_from_a_wanted_hacker.html*, April 27, 2006.

Klotz, Irene. "Apollo Astronaut Chats about UFOs, Alien Belief." Discovery Channel. *http://dsc.discovery.com/space/qa/alien-ufo-edgar-mitchell.html*, undated.

Lawhon, Loy. "Kecksburg." *www.ufoevidence.org/documents/doc1295.htm*, undated.

Lederberg, Joshua. "Engineering Viruses for Health or Warfare." *Washington Post*, August 16, 1970.

———. "Exobiology: Approaches to Life Beyond the Earth." *Science* 132, No. 3424, August 12, 1960.

———. "The Infamous Black Death May Return To Haunt Us." *Washington Post*, August 31, 1968.

———. "Mankind Had a Near Miss From a Mystery Pandemic." *Washington Post*, September 7, 1969.

———. "Medical Science, Infectious Disease and the Unity of Humankind." *The Journal of the American Medical Association*, August 5, 1988.

———. "A Treaty on Germ Warfare." *Washington Post*, September 24, 1966.

Lederberg, Joshua, with Carl Sagan and Elliott C. Levinthal. *Contamination of Mars*. Washington, D.C.: Smithsonian Institution, June 1967.

"Lincoln La Paz, 1897–1985." *www.math.ohio-state.edu/history/biographies/lapaz*, 1985.

Lindemann, Michael. "Interview With Gordon Cooper." The Institute for the Study of Contact with Non-Human Intelligence, February 16, 1996.

Link, Mae Mills. "Space Medicine in Project Mercury." NASA, 1965.

Lovgren, Stefan. "Far-Out Theory Ties SARS Origins to Comet." *http://news.nationalgeographic.com/news/2003/06/0603_030603_sarsspace.html*, June 3, 2003.

Luscombe, Richard. "NASA Told to Solve 'UFO Crash' X-File." *Observer*, November 11, 2007.

Mailer, Norman. *Of a Fire on the Moon*, London, UK: Little, Brown, 1970.

"Majestic 12, 1st Annual Report." *www.majesticdocuments.com/pdf/mj12_fifthannualreport.pdf*, 2010.

"Major General Theodore Cleveland Bedwell, Jr." The Official Website of the U.S. Air Force. *www.af.mil/information/bios/bio.asp?bioID=4642*, 1997.

Margerrison, Nick. "The Night Before." *Kerrang Radio*, July 23, 2008. *www.kerrangradio.co.uk/article.asp?id=1286863*.

McClatchey, Caroline. "How Gary McKinnon Became a Cause Celebre." *BBC News*. *http://news.bbc.co.uk/2/hi/uk_news/magazine/8181100.stm*, August 4, 2009.

McElwee, Jim. "The Vatican and NASA Alien Link." *www.unexplainable.net/artman/publish/article_9518.shtml*, May 26, 2008.

Mechanic, Michael. "Polis Report, Astro Nots." *San Jose Metro News*, January 1997. *www.metroactive.com/papers/metro/01.23.97/polis-rpt-9704.html*.

Mendenhall, Preston. "Amid sympathy, comspiracy theories." MSNBC. *www.msnbc.msn.com/id/3077629/ns/technology_and_science-space*, February 3, 2003.

Michael, Donald N. "Proposed Studies on the Implications of Peaceful Space Activities for Human Affairs." Brookings Institution, December 1960.

Ministry of Defense. UFO File. Reference number: AIR 2/17527, 1965.

Mitchell, Dr. Edgar, with Dwight Williams. *The Way of the Explorer*. Franklin Lakes, N.J.: New Page Books, 2008.

"Moon Dust." Report from U.S. Defense Attaché Office, La Paz, Bolivia, May 24, 1978, to Headquarters, U.S. Air Force, Washington, D.C.

"Moon Landing Conspiracy Theories." *http://en.wikipedia.org/wiki/Apollo_Moon_Landing_hoax_accusations*, 2010.

"Moon Movie." *www.moonmovie.com*, 2010.

"Moon Pyramid, The."*www.solarguard.com/tcreel1.htm*.

"Moon Shots 'Faked.'" *BBC News*. *http://news.bbc.co.uk/2/hi/world/monitoring/media_reports/1399132.stm*, June 21, 2001.

Moore, Waveney Ann. "Astronaut: We've had Visitors." *St. Petersburg Times*, February 18, 2004.

Moore, William L. "Crashed Saucers: Evidence in the Search for Proof." MUFON Symposium Proceedings, 1985.

Moore, William L. *The Roswell Investigation: New Evidence, New Conclusions*. Self-published, 1982.

Mullen, Leslie. "Alien Infection." *www.astrobio.net/exclusive/570/alien-infection*, August 25, 2003.

"NASA Employed Photo Artists to Airbrush out Apollo Anomalies." *www.ufos-aliens.co.uk/airbrush.htm*, undated.

NASA files on Joseph Perry. (Declassified under FOIA.) 1960.

"NASA to Search Files on '65 UFO Incident." MSNBC. *www.msnbc.msn.com/id/21494221*, October 26, 2007.

National Capital Area Skeptics. "Scientific Study of Unidentified Flying Objects." From Edward U. Condon, University of Colorado, 1968. *http://files.ncas.org/condon*, 1999.

Nevills, Amiko, "Remembering 'Gordo.'" NASA.gov. *www.nasa.gov/vision/space/features/remembering_gordo.html*, 2004.

"New NASA Legal Nightmare?" *www.realufos.net/2009/01/nasa-faces-legal-nightmare-credible.html*, January, 2009.

New Roswell: Kecksburg Exposed, The. SyFy Channel, 2003.

New York Times. Letter from Joshua Lederberg. July 13, 1969.

Newport, Frank. "Landing a Man on the Moon: The Public's View." *www.gallup.com/poll/3712/Landing-Man-Moon-Publics-View.aspx*, July 20, 1999.

Nowicki, Martin. "The Space Shuttle and its Replacement." *www.csa.com/discoveryguides/newshuttle/overview.php*, September 2004.

Oberg, James. "The Gemini 4 UFO," *UFO Report*, Fall 1981.

———. "In Search of Gordon Cooper's UFOs, 1983." *www.zipworld.com.au/~psmith/cooper.html*.

———. "Jim Oberg on the 'NASA Lawsuit Over Kecksburg UFO Documents.'" *www.jamesoberg.com/statement_nasa_kecksburg.pdf*, January 7, 2008.

———. "Massive Effort to Locate Cooper UFO Report." *www.jamesoberg.com/effort_locate_cooper_ufo_report.pdf*.

———. "Why did 'Gordo' Tell UFO Stories?" June 6, 2008, *www.jamesoberg.com/gordon_cooper2008comments.pdf*.

Patterson, William H., Jr. "Robert A. Heinlein, A Biography." *www.heinleinsociety.org/rah/biographies.html*, July 1999.

Peckham, Jeff. "Whistleblowers on UFOs and EBEs Support U.K. Hacker Gary McKinnon." *www.agoracosmopolitan.com/home/Frontpage/2009/01/23/03052.html*, January 23, 2009.

Percy, David S., and Mary Bennet. *Dark Moon: Apollo and the Whistle-Blowers*. Kempton, Ill.: Adventures Unlimited Press, 2001.

Pflock, Karl. *Roswell in Perspective*. Alexandria, Va.: Fund for UFO Research, 1994.

———. *The Day After Aztec*. Self-published, 2003.

Philips, Dr. Tony. "The Great Moon Hoax." *http://science.nasa.gov/science-news/science-at-nasa/2001/ast23feb_2*, February 2001.

Pippin, Ed. "Tom Corbett, Space Cadet." *www.solarguard.com/tchome.htm*, 1996.

Podesta, John. Statement delivered at the National Press Club, October 22, 2002.

Pope, Nick. *Open Skies, Closed Minds*. New York: Simon & Schuster, 1996.

———. *Operation Thunder Child*. New York: Simon & Schuster, 1999.

———. *The Uninvited*. New York: Simon & Schuster, 1997.

Posey, Bill, and Suzanne Kosmas. "Shuttle Flights Would Continue Under new Proposal." *Orlando Sentinel*. *http://blogs.orlandosentinel.com/news_politics/2010/03/shuttle-flights-would-continue-under-new-proposal.html*, March 3, 2010.

Poulsen, Kevin. "U.K. Hacker Gary McKinnon Plays the Asperger's Card." *Wired*. *www.wired.com/threatlevel/2008/08/uk-hacker-gary/*, August 28, 2008.

Press Officer. Home Office. Interview, December 20, 1996.

Printy, Tim. "Out of the Ashes." *http://home.comcast.net/~tprinty/UFO/ashes.htm*, July 2006.

"Project Mercury Overview." NASA.gov. *www-pao.ksc.nasa.gov/history/mercury/mercury-overview.htm*, 2000.

"RAAF Captures Flying Saucer on Ranch in Roswell Region." *Roswell Daily Record*, July 8, 1947.

Randle, Kevin D. *Crash: When UFOs Fall From the Sky*. Franklin Lakes, N.J.: New Page Books, 2010.

———. *Project Moon Dust: Beyond Roswell—Exposing the Government's Covert Investigations and Cover-ups*. New York: Harper, 1998.

Randle, Kevin D., and Donald R. Schmitt. *The Truth About the UFO Crash at Roswell*. New York: M. Evans, 1994.

———. *UFO Crash at Roswell*. New York: Avon, 1991.

Ranen, Aron. "Did we go? The Evidence is in!" *www.moonhoax.com/site/evidence.html*, 2002.

Redfern, Nick. *Celebrity Secrets: Government Files on the Rich and Famous*. New York: Simon & Schuster, 2007.

———. *Cosmic Crashes: The Incredible Story of the UFOs That Fell to Earth*. New York: Simon & Schuster, 1999.

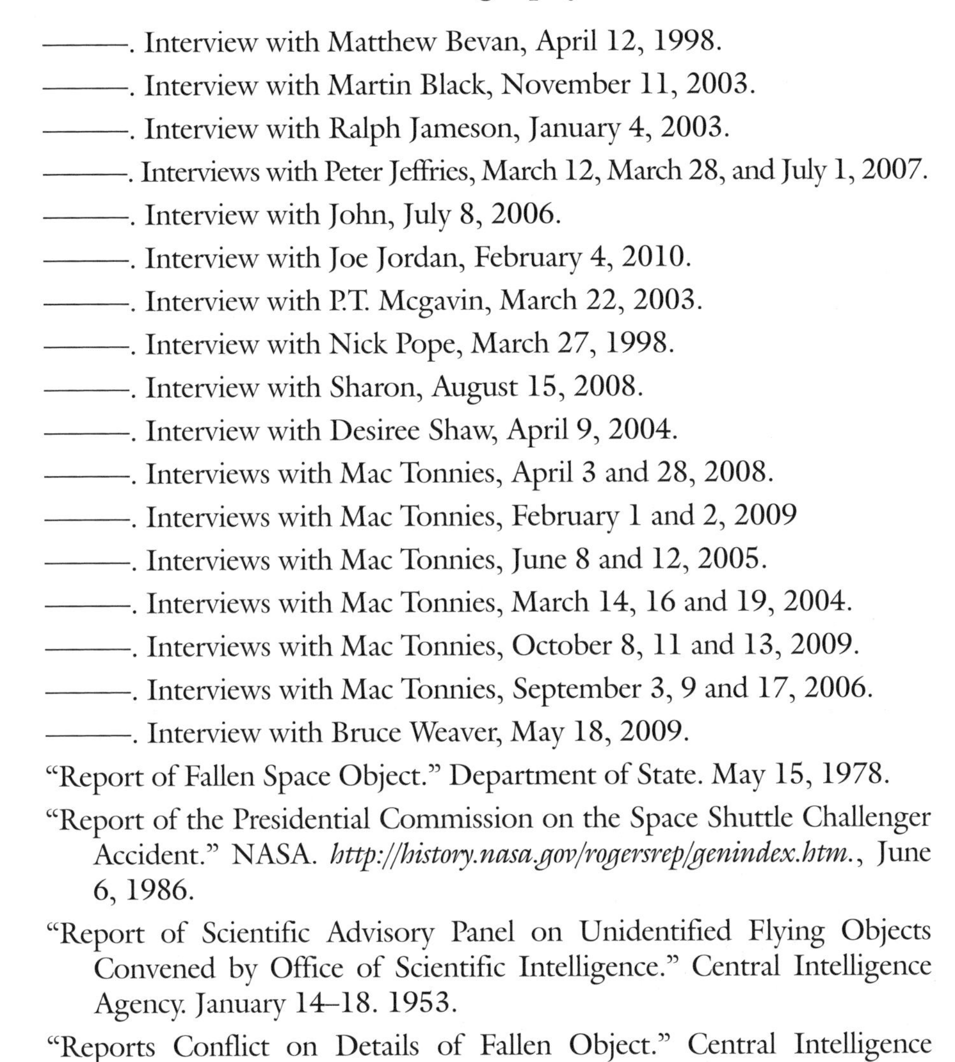

———. Interview with Matthew Bevan, April 12, 1998.

———. Interview with Martin Black, November 11, 2003.

———. Interview with Ralph Jameson, January 4, 2003.

———. Interviews with Peter Jeffries, March 12, March 28, and July 1, 2007.

———. Interview with John, July 8, 2006.

———. Interview with Joe Jordan, February 4, 2010.

———. Interview with P.T. Mcgavin, March 22, 2003.

———. Interview with Nick Pope, March 27, 1998.

———. Interview with Sharon, August 15, 2008.

———. Interview with Desiree Shaw, April 9, 2004.

———. Interviews with Mac Tonnies, April 3 and 28, 2008.

———. Interviews with Mac Tonnies, February 1 and 2, 2009

———. Interviews with Mac Tonnies, June 8 and 12, 2005.

———. Interviews with Mac Tonnies, March 14, 16 and 19, 2004.

———. Interviews with Mac Tonnies, October 8, 11 and 13, 2009.

———. Interviews with Mac Tonnies, September 3, 9 and 17, 2006.

———. Interview with Bruce Weaver, May 18, 2009.

"Report of Fallen Space Object." Department of State. May 15, 1978.

"Report of the Presidential Commission on the Space Shuttle Challenger Accident." NASA. *http://history.nasa.gov/rogersrep/genindex.htm.*, June 6, 1986.

"Report of Scientific Advisory Panel on Unidentified Flying Objects Convened by Office of Scientific Intelligence." Central Intelligence Agency. January 14–18. 1953.

"Reports Conflict on Details of Fallen Object." Central Intelligence Agency. May 16, 1978.

Rhodes, Tom. "UFOs: It's a Cover-Up." *Ottowa Citizen*, October 11, 1998.

Ronson, Jon. "Gary McKinnon: Pentagon hacker's worst nightmare comes true." *www.guardian.co.uk/world/2009/aug/01/gary-mckinnon-extradition-nightmare*, August 1, 2009.

———. "Jon Ronson Meets Hacker Gary McKinnon." *Guardian*, July 9, 2005.

Ryba, Jeanne. "Mercury: America's First Astronauts." *www.nasa.gov/mission_pages/mercury/missions/friendship7.html*, 2007.

———. "NASA Orbiter Fleet." *www.nasa.gov/centers/kennedy/shuttleoperations/orbiters/orbiterscol.html*, 2008.

———. "Space Shuttle." *www.nasa.gov/mission_pages/shuttle/main/index.html*, July 22, 2010.

Sagan, Carl. *Contact*. New York: Simon & Schuster, 1985.

Salter, Frank. "Astronaut Deke Slayton's Amazing UFO Encounter." *http://ufopartisan.blogspot.com/2009/10/astronaut-deke-slaytons-amazing-ufo.html*, October 2010.

"Sample Return Missions Scare Some Researchers." *www.space.com/searchforlife/planet_protection_000407.html*, April 9, 2000.

"Satellite Accidents With Radiation Hazards." Home Office, 1979.

Scalice, Daniella. "Vatican Hosts Study Week on Astrobiology, National Aeronautics and Space Administration." *http://astrobiology.nasa.gov/articles/vatican-hosts-study-week-on-astrobiology/*, November 17, 2009.

Schadewald, Robert J. "The Flat Out Truth: Earth Orbits? Moon Landings? A Fraud! Says This Prophet." *Science Digest*, July 1980.

"Searchers Fail to Find Object." *Tribune-Review*, December 10, 1965.

Second Look 1, No. 7, May 1979.

"Sergeant Karl Wolf: 'I've Seen Classified Photos of ET Moon Structures.'" Crystal Blue Mind. *http://newcrystalmind.com/2010/sergeant-karl-wolf-ive-seen-photos-of-et-moon-structures/*, April 27, 2010.

Sherwell, Philip. "Hacker Gary McKinnon will receive no pity, insists US." *Sunday Telegraph*, July 26, 2009.

Slack, James, and Michael Seamark. "An affront to British justice: Gary McKinnon extradition CAN be stopped, says Lib Dem QC." *Daily Mail*, May 31, 2010.

Slack, James, and Michael Seamark. "Gary McKinnon Extradition Stopped." *Daily Mail*, May 31, 2010. *www.dailymail.co.uk/news/article-1282765/Gary-McKinnon-extradition-stopped-says-LibDem-QC-Lord-Carlile.html#ixzz0qTo6ap4J*.

Slayton, Donald K., with Michael Cassutt. *Deke! An Autobiography*. New York: Forge Books, 1995.

"Somaliland President Egal Speaks on Mysterious Bomb Blast." CIA files, January 1996.

Space Information Officer, Royal Air Force Fylingdales, interview, December 20, 1996.

"Space Life Report Could Be Shock." *New York Times*, December 15, 1960.

"Space Shuttle Challenger Explosion." Federal Bureau of Investigation, *http://foia.fbi.gov/foiaindex/shuttle.htm*, 1986.

"Space Shuttle Columbia Disintegrates Upon Re-Entry." *PBS NewsHour*, February 1, 2003. *www.pbs.org/newshour/updates/columbia_02-01-03.html*.

"Spanish UFO Sightings." Department of Defense Intelligence Information Report, August 22, 1974.

Spence, Major Robert. Edwards Air Force Base, Office of Information Services, June 1957.

Stathopoulos, Vic. "Space Shuttle Columbia Disaster." *www.aerospaceguide.net/spaceshuttle/columbia_disaster.html*, 2010.

Streuli, Ted. "NASA Hires Writer to Debunk Apollo Theory." *www.galvestondailynews.com/report.lasso?wcd=5535*, October 31, 2002.

Stringfield, Leonard H. *Situation Red: The UFO Siege*. London, UK: Sphere Books, 1978.

———. *The UFO Crash/Retrieval Syndrome, Status Report II: New Sources, New Data*. Mutual UFO Network, January 1980.

———. *UFO Crash/Retrievals: Is the Cover Up Lifting? Status Report V*. Self-published, 1989.

"Tetrahedrons, Faces on Mars, Exploding Planets, Hyperdimensional Physics—and Tom Corbett, Space Cadet?! What Did They Know, and When Did They Know it?" *www.enterprisemission.com/corbett.htm*, undated.

Texas State University–San Marcos. "Texas State Research Sheds New Light On Panspermia." Press Release. *www.astrobiology.com/news/viewpr.html?pid=19104*, February 24, 2006.

"Tom Corbett, Space Cadet." *http://en.wikipedia.org/wiki/Tom_Corbett,_Space_Cadet, 2010.*

Tonnies, Mac. *After the Martian Apocalypse*. New York: Paraview Pocket Books, 2004.

"Transcript of Presidential Meeting in the Cabinet Room of the White House."*http://history.nasa.gov/JFK-Webbconv/pages/transcript.pdf*, November 21, 1962.

"Treaty on Principles Governing the Activities of States in the Exploration and Use of Outer Space, Including the Moon and Other Celestial Bodies, The." United Nations, Office for Outer Space Affairs, October 10, 1967.

Troy, Steve. "The 'Tom Corbett Lunar Pyramid'—Tying It All Together." *www.solarguard.com/tcreel1.htm*, 2000.

Truman, President Harry S. Press conference, April 4, 1950.

UFO 9, No. 3, 1994.

UFO 9, No. 5, 1994.

UFO Investigator, December 1960/January 1961.

UFO Universe 1, No. 3, Nov. 1988.

"UFOs and Extraterrestrials: A Problem for the Church?" *www.pufoin.com/pufoin_perspective/et_church.php*, July 6, 2001.

"UFOs over Brazil?" Department of State, May 21, 1986.

"Vatican and Alien Life, The." *www.ufoencounters.co.uk/the-vatican-and-alien-life.html*, 2010.

"Vatican Observatory." *http://vaticanobservatory.org*, 2010.

"Vatican Observatory: Welcome to the Vatican Advanced Technology Telescope." *http://vaticanobservatory.org/VATT/index.php*, 2010.

"Vatican: The Holy See." *www.vatican.va*, 2010.

Wade, Mark. "Ley." *www.astronautix.com/astros/ley.htm*, 1997.

Watson, Rob. "UFO Spotters Slam 'US Cover Up.'" *BBC News*. *http://news.bbc.co.uk/2/hi/americas/1322432.stm*, May 10, 2001.

"Welcome to the Astrobiology Science Conference!" *http://abscicon2006.arc.nasa.gov/abscicon2006.html*, 2005.

"Whistleblowers' Evidence of NASA UFO Fraud Might Kill UK Hacker Case." *Xenophilia*. *http://xenophilius.wordpress.com/2009/01/18/whistleblowers%E2%80%99-evidence-of-nasa-ufo-fraud-might-kill-uk-hacker-case*, January 18, 2009.

"White House 'Lost in Space' Scenarios." *The Smoking Gun*. *www.thesmokinggun.com/archive/0808051apollo1.html*, August 8, 2005.

Wilhelm, Allan R. "Grand Blanc Man Photographs Saucer." *Flint Journal*, March 28, 1960.

Willey, David. "Vatican Says Aliens Exist." *BBC News*. *http://news.bbc.co.uk/2/hi/europe/7399661.stm*, May 13, 2008.

Williams, Matthew. "A Close Encounter with Whistleblower Gary McKinnon." *5th Annual UFO Crash Retrieval Conference Proceedings*. Wood & Wood Enterprises, 2007.

"Willy Ley." *http://en.wikipedia.org/wiki/Willy_Ley*, 2010.

Wilson, Jim. "Space Shuttle Columbia and her Crew." NASA.gov. *www.nasa.gov/columbia/home/index.html*, August 23, 2006.

Wood, Ryan S. *Majic Eyes Only*. Broomfield, Colo.: Wood Enterprises, 2005.

Index

About the Author

Nick Redfern works full-time as an author, lecturer, and journalist. He writes about a wide range of unsolved mysteries, including Bigfoot, UFOs, the Loch Ness Monster, alien encounters, and government conspiracies. He writes regularly for *UFO Magazine*, *Fate*, *Fortean Times*, and *Paranormal Magazine*. His previous books include *Contactees*, *Memoirs of a Monster Hunter*, *There's Something in the Woods*, *On the Trail of the Saucer Spies*, and *Strange Secrets*. Nick Redfern has appeared on numerous television shows, including the BBC's *Out of this World*, History Channel's *Monster Quest* and *UFO Hunters*, the National Geographic Channel's *Paranatural*, MSNBC's *Countdown* with Keith Olbermann, and the SyFy Channel's *Proof Positive*. Nick co-hosts, with Raven Meindel, a weekly radio show on paranormal topics called *Exploring All Realms*. Nick Redfern lives in Arlington, Texas, with his wife, Dana. He can be contacted at *www.nickredfern.com*.